Department of the Environment

The Effects of Acid Deposition on Buildings and Building Materials

in the United Kingdom

Building Effects Review Group Report

London: Her Majesty's Stationery Office

620-191.3

Executive Summary

1. The main task of the United Kingdom Building Effects Review Group (BERG) has been to review the state of knowledge of the effects of acid deposition on buildings and building materials in the United Kingdom. To achieve this, five expert sub-groups were set up which dealt with air pollution modelling, site experience, laboratory tests, preservation techniques, and materials inventory. This interim report represents the first compilation of their findings.

2. As the basis for the report four main questions were addressed.

(i) *Is there a problem?*

The report concludes that current rates of weathering of stone on historic buildings and monuments appear to be higher than natural rates. Furthermore, weathering on all buildings is greater in urban than rural or remote areas for a number of materials including stone and some metals. Rates in coastal regions can be as high as in urban areas.

(ii) *How serious is the problem and is it getting better or worse?*

There is no unequivocal evidence that present rates of weathering of stone and most metals in the structure of historic buildings are significantly different from those in the recent past despite marked changes in emissions of the major air pollutants over the years. To a large extent the Group feels this reflects the complex nature of the interactions of pollutants, materials, and many other variables (eg turbulence, rain intensity) involved over long periods of time.

The pollution history of a material ('memory effect') especially of a porous substance such as stone, appears to be a major factor in determining current weathering rates and the Group identified this as an area in need of further understanding. Different materials also demonstrate differing sensitivities to air pollutants. Stones and metals interact strongly with sulphur dioxide, oxides of nitrogen and their reaction products. Concrete shows greatest reaction to carbon dioxide, while paints, plastics and organic material show greatest sensitivity to ozone and photo-oxidants.

(iii) *What types of building and which locations are most likely to be affected?*

Natural stone buildings are those most obviously at risk. Many of these are of historical interest. Modern buildings appear to be less at risk but greater numbers are involved. Two basic ways of assessing the stock of buildings and building materials that are at risk have been developed. These are:—

(a) the inventory/census approach which is more suitable for cultural buildings.

(b) a probability-distribution approach which is more suited to modern buildings containing repetitive units characteristic of modern production techniques.

(iv) *What trends can be observed in pollutant emissions and material damage?*

Average smoke and sulphur dioxide concentrations in urban areas of the United Kingdom have declined by factors of approximately ten and five respectively since the early 1950s. United Kingdom emissions of oxides of nitrogen have roughly doubled over the past 50 years. Since the 1950s, urban concentrations of sulphur dioxide have decreased faster than total emissions indicating an increased proportional contribution from high level emitters located outside urban areas. In a large number of areas, vehicles, particularly diesel powered, are now the major source of the smoke concentrations.

The Group feel unable at the present time to determine with any confidence relationships between pollutants and material damage relevant to real situations, including the effects of present and future reductions in emissions. They are also unable to undertake realistic economic assessments of the costs of damage and benefits likely to ensue from reduced emissions.

3. Attention is however drawn to the operational National Materials Exposure Programme. This

will provide information on the current rates of weathering for a range of building and constructional materials at 29 sites in the United Kingdom. This, together with the results of laboratory and field research work being currently undertaken in the United Kingdom, will better enable the Group to quantify the response to air pollution of a range of economically important building and construction materials within the next three to four years (1991-92).

4. In the meantime, an attempt is also being made to review the extent to which various materials are at risk in the pollution climates of the United Kingdom and what can be done to prolong material life in terms of preservation and protection.

5. BERG has identified a number of areas where further research or better data are needed. These include:—

— improved emission, deposition and micro-climate data especially for urban areas.

— better understanding of damage mechanisms leading to the derivation of realistic damage functions.

— quantification of the contribution of historical ('memory effect') and ambient pollution to current damage rates and evidence on whether a damage 'threshold' exists for buildings and construction materials.

— better methods of predicting damage on site specific, regional and national scales.

— extensions of the work on integrated costs of damage caused by pollution to individual buildings and methods of grossing up to urban, regional and national levels.

— further research into methods of preservation and protection, especially for ancient buildings and monuments, with a view to giving better advice on conservation and protection.

6. The Group does not consider that the evidence examined to date is adequate to enable its terms of reference to be met fully. It draws attention to further data which are likely to become available, and especially to the National Materials Exposure Programme (NMEP) (initial experiments to be completed by 1991) and recommends further report should be produced to take this new data into account in 1991-92.

United Kingdom Building Effects Review Group

L.H. Everett	Building Research Establishment, Garston	*Chairman*
T.M. Band	Scottish Development Department, Edinburgh	
P.A.T. Burman	Cathedrals Advisory Committee, London	
R.N. Butlin	Building Research Establishment, Garston	*Technical Secretary*
M.J. Cooke	British Coal Corporation, Stoke Orchard	
R.U. Cooke	University College London	
A.T. Coote	Building Research Establishment, Garston	*Minutes Secretary*
J. Eynon	Welsh School of Architecture, Cardiff (Replaced by J. Roberts)	
R.C. Haines	ECOTEC Research and Consulting Limited, Birmingham	
G.O. Lloyd	National Physical Laboratory, Teddington	
M.I. Manning	Central Electricity Research Laboratory, Leatherhead	
C.A. Price	English Heritage, London	
W.B. Whalley	The Queen's University of Belfast	
M.L. Williams	Warren Spring Laboratory, Stevenage	
R.B. Wilson	Department of the Environment, London	*Executive Secretary*
G.C. Wood	University of Manchester Institute of Science and Technology	
T.J.S. Yates	Building Research Establishment, Garston	*Editorial Assistant*

The authors accept responsibility for the contents of this report but the views expressed are their own and not necessarily those of the organisations to which they belong or the Department of the Environment.

Contents

List of Figures

List of Tables

Glossary

Acid deposition
Any form of deposition, dry or wet, arising primarily from the combustion products of fossil fuels or further reactions of these products.

Acid rain
Acid rain is defined as rainfall with a pH less than 5. In the absence of human activity natural long-term acidity would be of the order of pH 5.0. If saturated only with carbon dioxide at ambient concentrations the pH would be 5.6.

Activation
The degradation or corrosion of a previously stable metal surface by acid pollutants.

Black smoke
Those combustion particulates that can be measured by reflection techniques after absorption on a filter.

Bluff body
A body is aerodynamically 'bluff' when the flow streamlines do not follow the surface of the body, but detach from it leaving regions of separated flow and a wide trailing wake.

Cost benefit analysis
This allows the costs and benefits due to different factors to be isolated and each part expressed in real money terms.

Damage function
A function relating material degradation to environmental variables and pollutant concentrations.

Deposition resistance
The sum of aerodynamic, bluff-body and surface resistances, the total resistance to deposition of pollutants.

Deposition velocity
The concentration of a pollutant deposited on a surface per unit time divided by the ambient atmospheric concentration. Also defined as $1/r_t$, where r_t is the total resistance to deposition.

Dose response
The response of materials to pollutant concentrations.

Dry deposition
Pollutants reaching the ground in particulate form or as aerosols or gases.

Emitters
Sources of pollutant emission, usually divided into low (household), medium (factory, industrial) and high level (power stations) emitters.

Episode
A short period (minutes, hours or days) of elevated ambient pollution concentrations.

IRMA
Immission Rate Monitoring Apparatus – a passive monitor for measuring pollutant absorption.

Jacking
Mechanical expansion associated with crevice corrosion in metals.

Memory effect
The presence in building materials of pollutants which were absorbed in the past, but which can still cause degradation.

Micro Erosion Meter
Generic name for instruments used for measuring surface erosion (particularly on stone).

Molecular diffusion
The process by which pollutant gases are carried from the atmosphere to a surface.

Occult deposition
Pollutants intercepted in the form of aerosol droplets contained in mists, fogs and clouds.

Paretian
Economic theory which states that everything can have a value assigned to it.

Passivation
Formation of a tenacious oxide film on metals, often due to exposure to high levels of acid pollutants.

Preservation
Reconditioning or prolonging the life of building components that have already been damaged by weathering.

Protection
Improving the durability of new building materials which are going to be under atmospheric attack.

Restoration
The repair of buildings or building materials to their original form.

Reynold's Number
The Reynold's number is a nondimensional number which is the ratio of the inertia forces (mass × acceleration) to the viscous forces. Thus it is small when flow is slow.

Scale
Surface corrosion on metals.

Standing Building
An existing building, as distinct from ruins or archaeological remains.

Surface roughness index
Measurement of the 'waviness' and roughness components on the surface of a material.

Synergism
The enhancement of the degradation of building materials when two or more pollutants act together.

Wet deposition
Pollutants reaching the ground in rain and snow, or as occult deposition.

Abbreviations

ADMWG	Atmospheric Dispersion Modelling Working Group
AERE	Atomic Energy Research Establishment
ASTM (SP)	American Society for Testing and Materials (Special Paper)
AWRG	Acid Waters Review Group
BCC	British Coal Corporation
BISRA	British Iron and Steel Research Association
BRE	Building Research Establishment
BSI	British Standards Institution
CAC	Cathedrals Advisory Commission
CAPCIS	Corrosion and Protection Centre Industrial Service
CEC	Commission of the European Communities
CERL/CEGB	Central Electricity Research Laboratory/Central Electricity Generating Board
CRE	Coal Research Establishment (BCC)
DOE	Department of the Environment
ECOTEC	Environmental Research Consultants
ETSU	Energy Technology Support Unit (AERE Harwell)
GLC	Greater London Council
HMSO	Her Majesty's Stationery Office
ISO	Iron and Steel Organisation
IUAPPA	International Union of Air Pollution Prevention Associations
MAFF	Ministry of Agriculture, Fisheries and Food
NATO/CCMS	North Atlantic Treaty Organisation/Committee on the Challenges of Modern Society
NPL	National Physical Laboratory (Teddington)
NRPB	National Radiological Protection Board
NTIS	National Technical Information Service
OECD	Organisation for Economic Co-operation and Development
PAU	Programme Analysis Unit (AERE Harwell)
PORG	Photochemical Oxidants Review Group
PSA	Property Services Agency

RGAR	Review Group on Acid Rain
SORG	Stratospheric Oxidants Review Group
SSEB	South of Scotland Electricity Board
UCL	University College London
UEA	University of East Anglia
UMIST	University of Manchester Institute of Science and Technology
UNECE	United Nations Economic Commission for Europe
UWIST	University of Wales Institute of Science and Technology
WSL	Warren Spring Laboratory (Stevenage)

1 The Context of the Report

1.1 PURPOSE OF THE REPORT

The Building Effect Review Group (BERG) is one of a series of independent scientific committees set up by the United Kingdom Department of the Environment to give considered advice on the effects of acid deposition, in this case on buildings in the United Kingdom. Further impetus was provided by the requirements of the United Nations Economic Commission for Europe (UNECE) Convention on Long Range Transboundary Pollution and the House of Commons Environment Committee Report of September 1984, which drew particular attention to actual and potential effects on buildings and materials. The Government response to the Environment Committee Report was published in 1984 (DOE 1984). It outlined the action already being taken on many of the recommendations of the Report. The BERG was established as a consequence of this response to provide an independent review of the state of knowledge on the effect of acid deposition on buildings and building materials and to report to the Department of the Environment. The Review Group comprised a number of United Kingdom experts working in the field of air pollution and its effect on buildings and building materials.

1.2 TERMS OF REFERENCE

The terms of reference of the committee are:

  (i) To review the effects of acid deposition upon buildings and building materials in the United Kingdom;

  (ii) To orientate the review to the conditions currently prevailing in the United Kingdom;

  (iii) To undertake the production of an interim critical report in 1988;

  (iv) To identify gaps in knowledge and assess the need for further research.

1.3 SUB-GROUPS OF THE BUILDING EFFECTS REVIEW GROUP

The wide area of research covered by the terms of reference led to the formation of five sub-groups to study different aspects of the effect of acid deposition on buildings. The sub-groups dealt with the following areas:

  I. Air pollution modelling, with particular reference to the urban environment;

  II. Site experience (including information obtained from actual buildings and field trials); to include the development of a National Materials Exposure Programme;

  III. Intercomparison and validation of laboratory tests against field exposure;

  IV. Applied protection and preservation techniques;

  V. Materials inventory and evaluation of the building stock-at-risk.

In essence these groups have reviewed the various aspects of air pollutant damage on buildings including:

  — site and laboratory test data.

  — interrelationships between damage, pollutant concentrations and other environmental stresses.

  — protection and preservation systems.

  — assessment of materials and buildings at risk from pollutants (including economic appraisals).

The terms of reference of each sub-group are included in Appendix D.

1.4 THE STRUCTURE OF THE REPORT

The evidence submitted by the sub-groups of BERG has been used to answer four basic questions namely:

(i) Is there a problem?
(Chapter 2)

(ii) How serious is the problem and is it getting better or worse?
(Chapters 2, 6, 7, 8, 9)

(iii) What types of buildings and which locations are most likely to be effected?
(Chapters 3, 5, 6, 8)

(iv) What trends can be observed in pollutant emissions and material damage?
(Chapters 2, 4, 6)

The report is structured to present the evidence for acid deposition damage to buildings and building materials. It then shows the pathways from emission to deposition and degradation and discusses how materials can be protected or preserved. The later part of the report deals with the prediction of damage under changing pollution climates and discusses estimates of the costs likely to be incurred as a consequence of damage resulting from acid deposition.

The report also establishes a base for future studies, particularly by highlighting areas where present knowledge is deficient, and makes firm recommendations as to which areas most urgently require attention.

2 Is there a problem? – Evidence for damage to building materials by acid deposition

SUMMARY

The evidence put forward in this chapter shows all the features to be expected if atmospheric pollution is accelerating weathering and damage, but the evidence is not unequivocal that present rates of weathering on historical buildings are significantly different from those in the past. The evidence for metals shows a reduction in the rate of weathering for steel, but little to this effect for other metals. Nor does the available evidence indicate if 'safe' levels exist, or which atmospheric components are the major causes of damage – moisture, frost, salts, SO_2, NO_x, chloride, CO_2, etc.

However, since an effect does exist it is very important to assess the extent of this effect, both by measuring 'current' rates of weathering and damage and by assessing the stock at risk.

2.1 INTRODUCTION

The most important question to be answered in any discussion of the effect of acid deposition on building materials must be that posed by the title of this chapter – are buildings suffering damage and if so, how much is caused by acid deposition? In the past buildings were certainly disfigured by pollution including acid deposition. Now, 30 years after the beginning of major reductions in sulphur dioxide and smoke emissions it is important to examine the available evidence to determine if the situation has improved.

Evidence of pollutant effects on buildings can be drawn from two sources:

(i) Examination of standing buildings or samples taken from them, including the use of historic records, photographs, etc;

(ii) Atmospheric exposure of samples of building materials in the environment, with monitoring of environmental parameters.

The first obviously provides evidence obtained on actual buildings. However, it is difficult to isolate 'current' from 'historical' damage and also to distinguish damage caused by the present day atmosphere from that initiated by pollutants in the past but which are still able to cause damage by recrystallisation of reaction products from within the surface pores ('memory effect').

Site exposure trials on building materials provide evidence of the extent, if any, of current damage and degradation. However, it is difficult unless exposure is on an actual building, to reproduce environmental variables such as air turbulence, rainfall runoff, etc, nor can the environmental variables be controlled.

In the recent past the nature and rates of degradation of building materials have been monitored by a number of agencies. The different interests involved in these efforts has sometimes led to disparate results which have proved to be of limited value in the study of spatial and temporal variations of effects. The materials studied have sometimes also been limited in type to those of interest to each agency. The restricted nature of the studies has often made it difficult to define relationships between observed material changes, atmospheric conditions and weathering processes. However, the potentially high cost of reproducing atmospheric pollution makes it essential to obtain adequate data to determine the extent of the present problem and the precise relationships between spatial and temporal pollution levels, meteorological variables, and material damage.

The primary pollutants which are usually considered as leading to the degradation of building materials are sulphur dioxide, combustion particles ('black smoke'), and chlorides. Oxides of nitrogen (NO_x) may be involved either directly or indirectly but as Chapter 6 will show the evidence is as yet inconclusive. These primary pollutants can also lead to the formation of equally damaging secondary pollutants such as sulphates and nitrates. Many of these pollutant species occur naturally in the atmosphere and it is the excess of 'man-made' over 'natural' concentrations that is usually thought of as 'acid deposition'. The most prevalent atmosphere gas which can lead to 'natural' acid deposition is carbon dioxide. If saturated with dissolved CO_2 alone, rainwater has a pH of 5.6. The damage that this may cause in terms of limestone is poorly quantified (see 2.3.1.d) but there is evidence that CO_2 plays a role in the degradation of concrete (see 2.3.2). Also the effect of elevated levels of CO_2 on the acidity of rainfall or building materials, for example in an urban environment, are unknown.

Other factors that will lead to decay of some materials, whether pollutants are present or not, include meteorological variables such as wind speed and direction, temperature, relative humidity, and frequency and intensity of precipitation.

Any evidence obtained from past or present studies must be compared to 'background' damage ie that present in a natural unpolluted environment. However, 'background' damage is very difficult to estimate. It can be based on geomorphological evidence for long term erosion of rock out-crop surfaces, or by contemporary measurements in a geographical area where pollution is considered to be at a minimum.

Data for rates of 'natural weathering' are available for carboniferous limestone (Table 2.1). Data calculated from estimates of surface degradation since the start of the Holocene period ($\sim$ 10,000-12,000 years) suggests rates of 3-9 μm year^{-1} in Eire but 25-75 μm year^{-1} in Britain. These differences may reflect differences in the limestone, in precipitation, or in the methods used. The degradation rates based on rainwater runoff analysis (modern weathering) (23-88 μm year^{-1}) are higher than most of the surface degradation rates. Although this may reflect the durability of different limestones, it could also be interpreted as indicating that rates of weathering were higher in the recent past compared to the distant (pre-1700 AD) past.

The limited quantities of data on the effect of acid deposition on non-metals is a reflection of the amount of research undertaken in this area, it is not a measure of their relative importance.

2.2 EVIDENCE FROM STANDING BUILDINGS

The study of damage on standing buildings has generally been confined to qualitive assessments of damage over a number of years. This may take the form of comparison of photographs (Schaffer 1932; House of Commons 1984), or the time period between necessary maintenance work. Both of these techniques can give potentially valuable information. Quantitative studies are less frequent, but one relevant to the present discussion took place at St Paul's Cathedral, London (Butlin *et al* 1985) and is described below.

Table 2.1 Some 'natural' rates of carboniferous limestone weathering in the United Kingdom.

Area	Rock type	Rate (μm year^{-1})	Author	Method
Bare limestone outcrop analysis				
NW Yorks	Carboniferous limestone	40	Sweeting, 1966	1,2
C Claire, Eire	,,	9	High (pers comm)	1,2
C Claire, Eire	,,	3-4	Trudgill (pers comm)	1,2
S Wales	,,	25-75	Thomas, 1970	1,2
Limestone catchment runoff analysis				
Peak District	,,	75-83	Pitty (1968)	
Mendips	,,	40	Corbel (1959)	
Mendips	,,	23-88	Drew (1975)	
Mendips	,,	81	Smith & Atkinson (1970)	

Key (methods)

1-extrapolation from assumed glacial surface
2-water sampling

2.2.1 A Study of Portland Limestone at St Paul's Cathedral, London

The study was undertaken to determine the spatial variability of Portland stone weathering on St Paul's Cathedral, the historical rates of weathering, and the contemporary nature and rates of weathering (with monitoring of atmospheric conditions). The programme was complementary to the contemporary stone sample exposure and monitoring experiment organised by NATO/CCMS (see below).

(a) Rainfall and Runoff Monitoring

A photograph and the profile of the facade wall and balustrade adjacent to a North-East roof court on St Paul's are shown on Figures 2.1 and 2.2 together with the rainfall drainage zones. Rainfall drains from the coping of the balustrade onto the

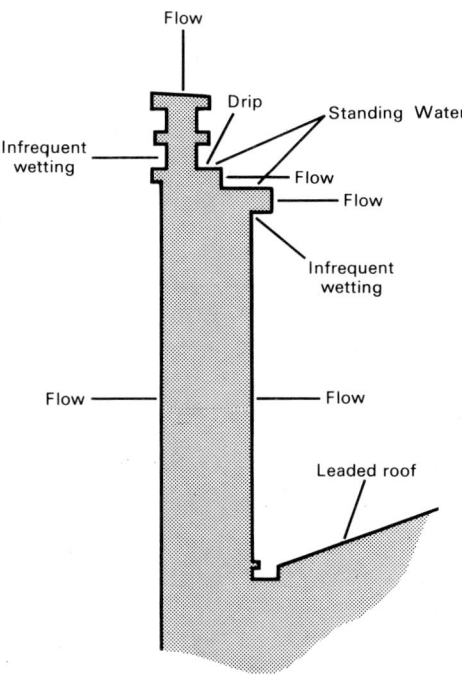

Figure 2.2   Profile of the facade wall of the North-East roof court, St Paul's Cathedral, London (wetting zones are indicated).

base and down to the roof court guttering system. Collection, sampling and analyses of the rainwater, runoff water at different points, and solids deposited were undertaken for five one-month periods in 1980-81. The following principal observations were made:

(i)   The mean hydrogen ion concentration of the rain water was equivalent to pH 3.8 to 4.9.

(ii)   As the runoff water passed over the stone surface the pH and total hardness (Ca^{2+}, Mg^{2+} cations) increased. The maxima occurred at the base of the facade wall.

(iii)   The concentrations of nitrate and sulphate in the rainwater were unrelated to the local concentrations of pollutants in the atmosphere. The concentration of nitrates and sulphates increased as water passed over the stone but the main source was the rainwater.

(iv)   The erosion sediment was mainly derived from the area first contacted by the rain.

Data from the St Paul's monitoring programme are given in Table 2.2. Within the 5 monitoring periods, conditions varied considerably and included some exceptional events, particularly two thunderstorms. The rain from one storm produced the lowest pH recorded (3.8) and this corresponded to the highest total hardness in solution in the

Table 2.2   Data for five monitoring periods St Paul's Cathedral, London.

Period	pH (Mean) (a) Rainwater (b) Flowing water	SO_4	NO_3	Total Hardness
		(mgl^{-1})	(mgl^{-1})	(mgl^{-1})
1	(a) 4.8	21.8	3.95	4.83
(16/10 - 12/11 1980)	(b) 6.6	23.6	4.80	154
2	(a) 4.6	15.8	4.46	7.0
(12/1 - 8/2 1981)	(b) 6.0	17.2	5.06	211
3	(a) 4.9	17.2	4.6	15.2
(9/3 - 5/4 1981)	(b) 6.4	19.9	5.4	96.8
4	(a) 4.6	14.3	4.6	13.5
(4/5 - 31/5 1981)	(b) 6.3	18.1	7.2	119.5
5 Day 1	(a) 3.8	3.8	18.3	24
(9/7/81)	(b) 6.0	4.4	19.5	610
Day 2	(a) 4.3	13.1	4.4	0
(22/7/81)	(b) 6.9	14.6	4.8	171

five periods of monitoring (610 mg 1^{-1}). On the same occasion the amount of erosion sediment was also the highest collected. The low pH was accompanied by high concentrations of nitrate (18.3 mg 1^{-1}) whereas sulphate concentrations (3.8 mg 1^{-1}) were lower than average.

(b) Surface Damage

It was possible from the study to establish a mean rate of limestone erosion within the balustrade catchment of 220 μm year^{-1}. This figure is much higher than those from micro-erosion meter measurements on the base of the balustrade and mortar exposure at base of the balustrade (139 μm year^{-1}). It is also much higher than the rate obtained from the differential erosion of stone and leads plugs (78 μm year^{-1}) (Figure 2.3) over the period 1727 to 1982 (Sharp et al 1982). The rate of erosion based on dissolution may be higher because it includes material dissolved from beneath the surface. Surface erosion measurements will not include this. The evidence presented here shows that damage is occurring and that the contemporary weathering rates may be greater than the long term historical ones. However, erosion and exposure also vary with position and aspect on the building (Sharp et al 1982) and with the exact nature and intensity of the rainfall.

The study at St Paul's Cathedral concentrated on rain-washed building surfaces. These often appear to be clean and in a better condition than sheltered surfaces. However, this is rarely true because the surface of the stone is being subjected to dissolution (confirmed by the increase in cation concentrations in rainwater as it passes over these surfaces). Sheltered surfaces tend to become disfigured by the formation of crusts, often blackened by the presence of soot. Several millimetres of crust can build up during prolonged exposure in a heavily polluted atmosphere (see Chapter 6 for further discussion).

One outcome of these observations relating the behaviour of sheltered and rainwashed surfaces observed on standing buildings (see below) has been the design of exposure racks for site exposure trials involving the pairing of identical samples in exposed and sheltered conditions.

2.3 EVIDENCE FROM SITE EXPOSURE TRIALS

2.3.1 Stone

Site exposure trials of stone examples have taken place for over 60 years (Schaffer 1932; Butlin et al 1985). The earlier studies yielded estimates of damage to different stones at a number of sites, but only in relative not absolute terms. More recent trials have included measurement of environmental parameters, including pollutant levels.

(a) The Garston – Westminster Study

The Garston – Westminster Portland stone exposure study took place between 1955-1965 and

the results allowed estimates of past weathering rates to be made.

Blocks of Portland stone (100 × 50 × 50 mm) were exposed at the Building Research Station, Garston, Herts and at Westminster (Whitehall). Series A were exposed from 1955 to 1965, Series B from 1957 to 1965. Although all the samples were of Portland stone, the measured porosity and saturation coefficients (measures of durability) differed significantly between the two series. The weight loss of each block was measured at approximately two yearly intervals. Atmospheric concentrations of SO_2 were measured using the lead dioxide candle method. The results, shown in Figure 2.4 (Jaynes 1985) and Table 2.3, indicated

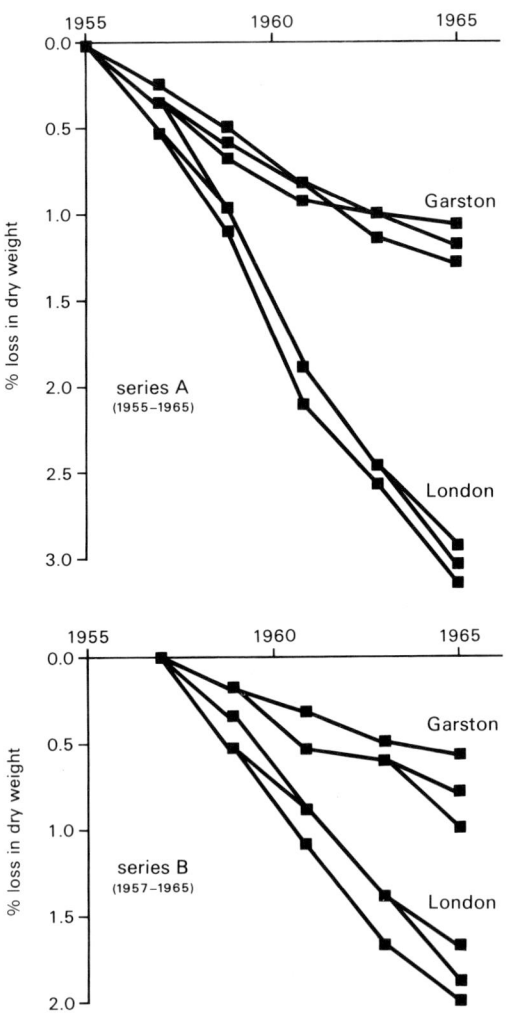

Figure 2.4 Results from exposure trials of stone samples at Garston, Hertfordshire, and Whitehall, Westminster.

greater weight loss where average SO_2 concentrations were higher, but the rate of decay was not found to be directly proportional to the pollution level as indicated by the measurement technique.

Table 2.3 Annual average sulphur dioxide pollution measurements (based on daily readings).

Twelve month period ending:	Garston		Whitehall	
	Summer	Winter	Summer	Winter
31.2.1955	13	41	132	275
31.3.1956	14	43	136	288
31.3.1957	16	43	130	272
31.3.1958	14	49	130	277
31.3.1961	20	51	119	239
31.3.1962	20	46	114	—
31.3.1965	16	43	136	244
31.3.1966	19	49	109	244

All units are μm^{-3} and are calculated from 'lead candle' measurements assuming a deposition velocity of 17 mm s^{-1}.

The weight losses are equivalent to minimum surface weathering rates of 23-33 μm year^{-1} at Westminster and 10-13 μm year^{-1} at Garston. However, if the upper surface was preferentially degraded the rates could be up to six times higher (60-198 μm year^{-1}).

(b) The NATO/CCMS Programme on Stone Decay 1980-1982

A programme of measurement of air pollutants and their effect on two types of stone, Baumberg sandstone and Krensheim Muschelkalk (fossiliferous limestone), was carried out at 27 sites in Europe and North America (Table 2.4) over a period of two years (1980-1982) (NATO 1985) An Immission Rate Monitoring Apparatus device (IRMA, a passive monitor for measuring pollutant absorption) plus two carousels of stone-tablets (60 × 60 × 2-3 mm) (one sheltered and only open to dry deposition; the other open to wet and dry deposition) were located at each site. Measurements of pollutant uptake were made at each site on a two weekly basis. At the end of the two year period, measurement of the overall

Table 2.4 Site locations in the NATO/CCMS Study.

Sites	
Great Britain	St Paul's Cathedral, London; Westminster Abbey, London; Dunstaffnage Castle (Scotland)
Netherlands	Lelystad; Den Burg, Texel; Great Church, Haarlem; Old Church, Delft
Italy	Ravenna; Pisa Cathedral; Basilica S. Marco, Venice; Rome
Norway	Bergen
Sweden	Stockholm; Floda
France	Douai; La Rochelle; Strasbourg
USA	New York Custom House, then NPS Federal Hall
Greece	Acropolis, Athens; Eleusis; Hadrians Gate, Athens
Federal Republic	Cologne; Ulm; Dortmund - Bovinghausen; Haus Alst

weight losses of stone specimens was undertaken centrally in Germany together with destructive chemical analysis of the stone samples.

Atmospheric Pollutant Concentrations: Rates of uptake were determined in the IRMA equipment in mg m^{-2} day^{-1}. The annual average rates for 1980/81 for SO$_2$ ranged from London (St Paul's) (128 mg m^{-2} day^{-1}) down to (Floda) (6.5) and Dunstaffnage (12). For 1981/82, Rome was the highest (123) with London (Westminster Abbey) (90) down to Floda (8) and Dunstaffnage (9). The highest chloride values were found for those sites nearest the coast, values ranged from 93 at Den Burg down to 2 at Floda, London had values of 6-8. Values for SO$_2$ were generally much higher in winter than in summer.

Levels of NO$_2$ over the 2 year span were highest for Haarlem and New York (12) Athens (11), and Cologne (10), often with considerable variation from year to year. One third of the sites, including London, Venice, Stockholm, Athens, Dortmund, produced values in the range 1-4 mg m^{-2} day^{-1},

Analysis of Samples: Losses in weight were noted for unprotected stone from all but one of the sites. The losses were higher for Baumberg stone than Muschelkalk and higher for conurbations than elsewhere. Increases in weight were noted for most of the sheltered specimens with weight increases relating generally to pollution levels. Samples of Baumberg stone and Muschelkalk at St Paul's Cathedral, London lost 8% and 7% in weight (exposed) and gained 2% and 0.6% (sheltered) respectively. The weight loss is equivalent to a minimum surface weathering rate of 33-55 μm year^{-1}.

The pollutant contents of exposed samples were about the same as the background levels in new unweathered specimens of stone, probably because the pollutants were washed out by rainwater. Some elevated levels (London, Athens) were attributed to dust layers on the surface. Appreciable concentrations of all pollutants were found on sheltered samples, again with the highest values for conurbations.

Some correlations were found to exist between the pollutant contents of the stone and the concentrations measured by the IRMA device. The highest correlations were for SO$_2$ and were obtained for Baumberg with a power function and Muschelkalk with a linear function.

The concentrations of pollutants deposited on specimens or the IRMA device showed a negative correlation with the rate of absorption. This suggests that degradation may depend on mechanical flow effects in the atmosphere around the stone. It was felt at the end of the first phase of work that some limited success had been achieved but that other parameters would have to be taken into account (eg meteorological measurements) before a more complete cause-effect analysis was possible. It should also be remembered that the IRMA device is dependent on approach velocity and to some extent on the aerodynamics around a building. Adequate data were not available in these experiments to calibrate the device against standard methods of measurements of pollutant concentrations. Further analysis using 'rain-days', or surrogates for this, gave correlations with SO$_2$ on an empirical basis.

(c) Spatial Variability of Building Stone Weathering in Southern Britain

This programme was designed to determine whether the higher concentrations of atmospheric pollutants in urban areas lead to accelerated weathering of limestones (Butlin et al 1985). Twenty-five sites were chosen in London and the South-East of England (Figure 2.5), to provide a representative sample of city, town, country and coastal conditions.

Pairs of freely-rotating carousels containing 6 tablets of each of two limestone types (Portland, a durable building stone and Monks Park, a less durable stone) were placed at each site. The condition of the weathered stone tablets was monitored by weight loss, changes in surface morphology and chemical analyses. The weight loss for the sheltered samples was measured after leaching with distilled water. The period of exposure was two years, a period long enough to produce significant weathering of the stone samples.

Weight Loss: Regional variations of weathering were found (Figure 2.5) with rates increasing from rural to provincial urban areas and to London (~ 25% greater than rural). The rural-urban increase may be greater elsewhere in England where the rural areas are relatively less polluted. The weight loss of stone at St Paul's Cathedral was the second highest of all the sites after Bletchley. The leached loss of sheltered carousel samples was approaching half that of the exposed carousel samples, suggesting that the processes by which the sheltered samples acquired soluble material were partly related to the processes causing weight loss in exposed samples. Dry deposition of pollutant gases, especially by sulphation, must be the major source of soluble reactants (formed under humid conditions) for the sheltered samples and consequently the same

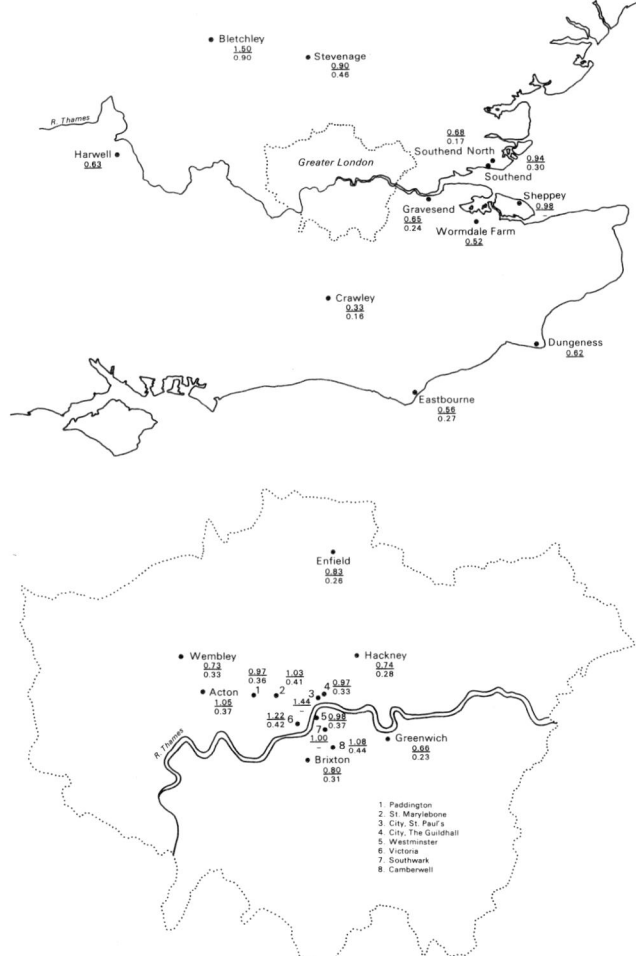

Figure 2.5 Location of sites and percentage weight losses (exposed/sheltered) for Portland stone samples at sites in South-East England (after Jaynes 1985).

process may account for up to 40% of exposed sample weight loss.

The remainder of the weight loss on exposed samples is likely to be caused by wet deposition of pollutants, with the rain acting as a solvent for both gypsum and calcite and also as an acid.

Surface Morphology: Surface roughness was determined for the Portland stones after the two year programme was complete. Surface roughness indices increased with weathering for exposed and sheltered samples. The roughness indices of exposed samples correlated with weight loss, and the roughness of sheltered samples correlated with leached weight loss. In general, exposed samples become rougher with time than sheltered samples.

Chemical Analyses: Sulphate, nitrate and chloride levels were determined for sheltered samples and all increased during the study. Chloride concentrations showed a peak at coastal sites (marine aerosol sources) and also a significant increase between rural and urban sites.

Increases in sulphate levels with weathering were an order of magnitude greater than the increases for chloride or nitrate. Sulphate concentrations also increased in exposed samples suggesting the formation of gypsum below the sample surface. Stones from central London were found to have a greater concentration of sulphate than those from rural sites. Sulphate concentrations in sheltered stones were correlated with weight loss for the exposed samples at the same site, suggesting a relationship between weight loss and the process of uptake of SO_2 by dry samples. However, no significant correlation was observed between atmospheric SO_2 and weight loss.

None of the weathering indices were significantly correlated with rainfall, frequency of frost or NO_x concentrations.

(d) The Role of Carbon Dioxide in the Dissolution of Calcium from Limestone Specimens

Martin and Barber (1987) used results from the NATO/CCMS Programme (above, 2.3.1.b) and the 'run-off' results from St Paul's Cathedral (Butlin *et al* 1985, above 2.2.1) to suggest that part of the loss of weight and calcium ions could be attributed to carbon dioxide. Consequently Martin and Barber (1987) set up laboratory and field experiments in an attempt to quantify the effect of CO_2, SO_2, NO_x, sea salts, and acidity in rainwater on the dissolution of calcium and other ions from limestone species. They conclude that in their particular experiment, at an atmospheric concentration of 30 μgm^{-3} of SO_2, the relative contributions to limestone loss were bicarbonate 55% (from CO_2), acid anions (from the original rainfall) 24%, and sulphate (from atmospheric SO_2) 21%. This indicates that CO_2 may make a significant contribution to degradation of stone. However, this study involved only two small blocks of stone and there are as yet no other studies to validate these conclusions.

2.3.2 Other Non-Metallic Materials

There are very few records of site exposure trials that have included non-metallic materials other than stone.

Samples of concrete have been exposed for 20 years at a heavily polluted site at Beckton in East London and at Garston, Hertfordshire. Those at Beckton show a marked erosion of the surfaces with the flint aggregate left standing proud. This erosion is thought to be a surface effect only and not to play so important a part as carbonation which occurs deep in the pore structure. It would take a long time for the aggregate to be etched out of the matrix and this would not occur within the lifetime of a reinforced concrete building.

2.3.3 Metals

(a) Introduction

Atmospheric corrosion of metals has been shown to play an important part in the economy of industrialised societies (Mansfield 1980; 1982a) and has been the subject of many descriptions, compilations and reviews (Evans 1960; Rozenfeld 1972; Barton 1973; Ailor 1982; Carter 1982).

Atmospheric testing in the United Kingdom dates back to the nineteenth century (Parker 1881). By the 1930s the work of Evans (1960) and Vernon (1943) had led to a good understanding of the broad general principals that affect rates of atmospheric attack on metals, and a number of comparative trials (Burgess & Aston 1913; Buck 1913; Friend 1929a; 1929b; Hudson 1929) had been organised in different atmospheres (Table 2.5). In general, little quantitive information on atmospheric pollution associated with each site was available, sites being designated as 'rural', 'marine', 'urban', 'industrial' etc.

The systematic series of trials undertaken by the British Iron and Steel Research Association from the 1930s (BISRA 1938; BISRA 1959) did include some measurements of rainfall constituents and, later, of atmospheric SO_2. In general, relationships between corrosion and pollutant levels were not derivable from these trials although the work of Hudson and Stanners (1953) on zinc has been analysed in terms of SO_2 concentration.

These trials succeeded in their primary aims of providing information for materials selection and protective measures to be taken. During the last 30 years, improved pollution measuring capabilities have been developed and urban pollution levels have changed dramatically. However, few extensive or systematic corrosion surveys in the United Kingdom have taken advantage of these changes. Thus the development of predictive functions has occurred only sporadically and there is a general lack of information on contemporary corrosion rates.

Metals and other building materials may be aesthically damaged by the environment even if structural damage is not present. This applies in particular to copper and its alloys, and

Table 2.5 United Kingdom atmospheric corrosion trials on metals.

Metal	Reference	Comments
Aluminium and Alloys	Liddiard and Whittaker (1961)	Depth of attack
	Latimer and Booth (1963)	Mechanical properties; 4 sites
	McGeary *et al* (1968)	Depth of attack
	Booth and Goddard (1965)	Depth of attack; Mechanical properties
	Carter (1968)	Weight loss; depth of attack; mechanical properties; 5 sites
	Ailor (1969)	Weight loss; depth of attack; mechanical properties
	Hailstone (1969)	Weight loss
	Ailor (1974)	Weight loss; depth of attack, mechanical properties
	Lifka (1977)	Weight loss; depth of attack; mechanical properties
	Skerrey (1982)	Weight loss; depth of attack; mechanical properties
Copper and Alloys	Vernon (1923; 1927)	Visual; chemical analysis; weight loss; indoor and urban
	Friend (1929a)	Visual; weight increase, urban
	Hudson (1929; 1930; 1935)	Visual; weight loss, electrical resistance; ultimate tensile strength; rural; urban; marine
	Scholes and Jacob (1970)	Visual; chemical analysis; electron diffraction; weight loss, metallography; yield stress; ultimate tensile strength; industrial; marine
Lead and Alloys	Friend (1929a)	5 sites
	Clarke and Bradshaw (1953)	
	Hudson and Stanners (1953a)	
Iron and Steel	Parker (1881)	Steel discs
	Burgess and Aston (1913)	Iron alloyed with various elements
	Buck (1913)	Effect of copper content
	Friend (1929b)	Cast iron; wrought iron; carbon steel; nickel-chromium low alloy steels
	BISRA (1938)	Ingot iron; wrought iron, mild steel; cooper steels; various finishes; several sites
	BISRA (1959)	As above; rainwater analysis;
	Hudson and Stanners (1953b)	As above; SO_2 correlation; Cl^- effects
	Higgins (1956)	SO_2 correlation
	Chandler and Kilcullen (1968)	Mild steel; 22 sites around Sheffield; correlation with SO_2 account for 50% of corrosion
	Truman (1979)	Stainless steels
	Saunders (1981)	Mild steel; stainless steel; 8 sites with environmental monitoring; fumigation site; no published results
	McKenzie (1982)	Weathering steels; exposed; sheltered; several sites; SO_2; Cl^-; temp; RH; rainfall amount
Tin	Kenworthy (1934)	Indoor; outdoor
	Hudson and Banfield (1946)	Coatings on steel
	Hudson and Stanners (1953a)	
	Britton (1976)	
Zinc	Hudson and Banfield (1947)	Zinc and zinc coatings
	Hudson and Stanners (1953b)	Correlation with SO_2
	BISRA (1959)	
	Britton and Angeles (1951)	Coatings
	Clarke and Bradshaw (1953)	Coatings
	Gilbert (1953)	2% Cu
	Ambler (1960)	Coatings
	Layton (1965)	Coatings
	Hutchins and McKenzie (1973)	
	Shaw (1978)	Zinc cans; ongoing; corrosion map
	Saunders (1981)	Zinc cans; galvanized panels; 8 sites with environmental monitoring; fumigation site; no published results
Magnesium	Shrier (1976)	Industrial
Cadmium	BISRA (1959)	Coated steels
Nickel and Alloys	Hudson (1929)	Nickel; 80/20 Ni-Cr; 70/30 Ni-Cu
	Evans (1969)	Industrial
	Carter (1982)	

aluminium. For example, stone or brick can be stained by green copper corrosion products. Aluminium, when anodised or dyed, can be discoloured by atmospheric exposure.

(b) Iron and Steel

Iron and steel have been submitted to atmospheric corrosion trials since the nineteenth century. The corrosion is primarily due to oxygen and moisture, but is accelerated by contaminants such as SO_2, particulates, and chlorides. Additions of copper, nickel and chromium reduce the corrosion susceptibility of steel.

It may be seen from Table 2.5 that a number of studies of the atmospheric corrosion of iron and steel have been undertaken in the United Kingdom. In very few, however, have any functional relationships between damage and pollutants and meteorological conditions (damage functions) been determined. More extensive investigations of this nature have been undertaken abroad particular examples being those of Haynie and Upham (1971), Barton and Czerny (1980), Barton (1973), Mikailovsky et al (1980) and Hakkarainen and Ylasaari (1982). The relationships derived by these workers differ in form, in the variables used and, in the particular case of 'time of wetness', in definition.

Observed corrosion rates vary greatly with the nature of additives. A typical range from the BISRA (1938; 1959) trials for ingot iron is 48 μm year^{-1} (at Godalming, Surrey) to 173 μm year^{-1} (at Derby). A 0.3% copper steel exhibited rates between 36 μm year^{-1} and 109 μm year^{-1} in the same trials.

Data obtained by the British Steel Corporation at Stratford in East London show that rates of corrosion of copper-bearing steel were 62.5 μm year^{-1} in 1979-80 and 37.6 μm year^{-1} in 1983-84, in line with a considerable fall in SO_2 levels (measured by the lead dioxide candle method). The latter figure is among the lowest recorded for this type of steel and may be compared with rates of 100 μm year^{-1} or more at industrial sites in the period 1933-1960.

(c) Zinc and its Alloys

Zinc corrodes in the atmosphere at a relatively slow and essentially constant rate. Exposure to SO_2 increases the degradation rate. Pure zinc has been widely used as a standard sample to determine the corrosivity of atmospheres (Shaw 1978; MAFF 1986) and in the form of galvanising has also been included in a number of trials.

Typical corrosion rates observed in the BISRA (1938, 1959) trials are 1 μm year^{-1} (Godalming, Surrey) to 10 μm year^{-1} (Frodingham, Lincolnshire). As with iron and steel, there have been a number of surveys of zinc or galvanising corrosion for which functional relationships with atmospheric parameters have been derived (see Chapter 7).

(d) Copper and it Alloys

Copper and its alloys are highly resistant to atmospheric corrosion due to the formation of a green surface film consisting principally of cuprous oxide together with basic salts. Corrosion attack is uniform over the surface. Typical penetration rates range from 0.2 to 0.6 μm year^{-1} in rural environments to 0.9 to 2.2 μm year^{-1} in industrial areas (Mattson & Holm 1968).

(e) Lead and its Alloys

Lead and its alloys are highly resistant to corrosion in polluted atmospheres due to the formation of protective oxides, sulphides, sulphates and carbonates. Performance in chloride contaminated atmospheres may be slightly impaired.

(f) Aluminium and its Alloys

Most aluminium alloys have excellent resistance to atmospheric corrosion through the formation of a thin, tenacious oxide film. The surface becomes roughened because of shallow pitting and accumulations of corrosion products but there is no general thinning of the metal except after extended exposures in polluted atmospheres. In an industrial environment, over a six year period, McGeary et al (1968), observed an average corrosion rate of 2 to 5 μm year^{-1}. However, a number of small pits reached a depth of 250 to 500 μm over the same period.

(g) Tin and its Alloys

When exposed to atmospheric environments tin corrodes only slowly. The attack is uniform and almost constant with time, leaving a grey product which is mainly stannous oxide. Pollutants, in general, have little effect. In many years of trials by the American Society for Testing Materials (ASTM) in Britain the following corrosion rates have been observed (Britton 1976):

Industrial 0.125-0.175 μm year^{-1}
Coastal   0.175-0.275 μm year^{-1}
Rural   0.05 μm year^{-1}

(h) Magnesium and its Alloys

Magnesium alloys are essentially immune from corrosion in the atmosphere at relative humidities below about 60% but corrosion spots do develop when the humidity approaches 100%. Corrosion is generally linear with time and rates of up to 76 μm year^{-1} have been observed in the United Kingdom in an industrial environment (Shrier 1976).

(i) Cadmium

Cadmium has a similar corrosion performance to that of zinc but its use tends to be confined to that of protective coating applied to steel by electrodeposition.

(j) Nickel and its Alloys

Nickel is highly resistant to atmospheric corrosion though tarnishing occurs on exposure to SO_2 at high relative humidities. Corrosion of nickel is linear but alloys have been observed to exhibit corrosion rates which decrease with time. Nickel is widely used as a coating often with an overcoating of chromium or precious metal to avoid tarnishing. The performance of nickel-chromium coating systems has been extensively studied. Stress corrosion of some nickel alloys has been observed in high nitrate environments.

Average penetration rates for pure nickel have been observed in the range 0.25 μm year^{-1} (rural) to 1.75 μm year^{-1} (industrial) in the USA, but higher rates have been reported for the United Kingdom (Carter 1982).

2.4 CONCLUSIONS

There is no doubt that building materials are adversely affected by weathering. The critical question is whether the presence of pollutants in the atmosphere (and rainwater) accelerates the rate of damage. If this is true, then the following features should be present in the data described above:

(i) An increase in the rate of weathering, relative to the 'natural' background, where the pollutant levels are higher (eg urban relative to rural areas);

(ii) Evidence that atmospheric pollutants can damage building materials;

(iii) Evidence for uptake, by building materials, of pollutants from the atmosphere.

Comparison of 'natural' and 'contemporary' corrosion rates is difficult because of the limited amount of data and because the measurements are on different materials in different geographical locations, and often by different methods.

The rates of weathering (estimated from weight losses) for stone samples measured by exposure at Garston, Westminster and St Paul's Cathedral appear to fall within the range of 'natural' rates. However, measurements made on historical building components are very much higher than 'natural' rates and show a variation with aspect and position. The pollutants involved are generally SO_2, oxides of nitrogen (NO_x), carbon dioxide and particulates. In coastal areas, chloride plays an important part.

Higher rates of weathering of new stone were measured at urban sites compared to rural sites in the study of stone in South-East England (Jaynes 1985; Butlin et al 1985) and in the NATO/CCMS study. However, in the latter, rates in coastal regions were as high as in urban areas (NATO 1985). A similar rural/urban increase is recorded for iron, copper and tin.

Chemical changes in building materials were observed on new stone samples exposed in South-East England (Jaynes 1985; Butlin et al 1985). Sulphate concentrations increased on both sheltered and exposed samples. Nitrate and chloride concentrations also increased but less dramatically. No direct correlation between atmospheric SO_2 and weight loss was observed. The presence of sulphates on metals (particularly iron and zinc) is also indicative of the role of pollutants in corrosion.

The evidence put forward in this chapter shows all of the features to be expected if atmospheric pollution is accelerating weathering and damage. However, the evidence is not yet unequivocal that present rates of weathering on historical buildings are significantly different from those in the past. The evidence for metals shows a reduction in the rate of weathering for steel, but little to this effect for other metals. Nor does the available evidence indicate if 'safe' levels exist, or which atmospheric components are the major causes of damage - moisture, frost, salts, SO_2, NO_x, chloride, CO_2, etc.

However, since an effect does exist it is very important to assess the extent of the effect, both by measuring 'current' rates of weathering and damage and by assessing the stock at risk.

3 Which buildings and materials are at risk, and how can the total 'stock-at-risk' be estimated?

SUMMARY

Estimating the number of buildings and the amount of material at risk from acid deposition is clearly a difficult and complex process. However, it is essential for any estimate of the cost of damage, and consequently for any form of cost-benefit analysis. The two main alternative methods each have their advantages and disadvantages.

As the cultural and aesthetic significance of historic buildings and monuments of national and international importance may be much greater than their cost of replacement, the Inventory/Census method would appear to be the most appropriate. The Probability – Distribution approach appears to be the more successful and more suitable for modern buildings.

3.1 INTRODUCTION

In order to be able to assess the damage caused to buildings and building materials by acid deposition four questions must be considered:

   (i)   Which materials are affected?;

   (ii)   Which parts of a building are most likely to be affected?;

   (iii)   What type of buildings are likely to be affected?

   (iv)   How many buildings are likely to be affected?

The answers to these questions are important for estimating the cost of damage caused by acid deposition and, consequently, in considering the cost-effectiveness of any actions taken to reduce pollution and acid deposition.

3.2 WHICH MATERIALS ARE AFFECTED?

The previous chapter indicated some of the materials that are considered to be at risk. However, it is important to distinguish between materials that are very susceptible to pollutant damage and those that are less susceptible but more often used. For example, concrete is only slightly damaged by acid deposition but the huge quantities involved could increase the financial importance of any damage. This is considered further in the discussion on cost-effectiveness in Chapter 8.

A list of the main building materials that are affected is given in Table 3.1 and an indication of whether their main uses are structural, as accessories, or decorative. The table shows metals and calcareous stones (limestones, magnesian limestones and calcarous sandstones) to be the worst affected by acid deposition. Other materials are damaged by 'natural' processes but these are not worsened by acid deposition. However, one exception to the latter group is probably medieval glass. This is known to be damaged by moisture and decay products including sulphate, but the role of the sulphate in this decay is not clear (see Chapter 6).

It is also possible that organic materials (some of which are filled with inorganic compounds), brickwork, cement, concrete and non-calcareous slates might be affected in some way. The durability of such materials in the absence of pollutants is a necessary consideration in any assessment of additional effects caused by acid deposition.

Some materials exhibit decay phenomena peculiar to their group, an example being the cavernous decay of magnesian limestones. Others exhibit decay characteristics that do not relate directly to current pollutants or climatic conditions, but to past conditions (sometimes termed the 'memory effect'). Organic growths (algae, lichens, and

Building materials		External elements		Secondary roof elements					Roofs			Secondary wall elements													Walls			
		Outbuildings	Curtilage wall/fencing	Accessories on roof	Roof drainage	Dormers	Roof lights	Chimneys	Finishes to roof	Flat	Pitched	Lintels	Sills	Grilles	Doors	Shop fronts	Windows	Stairs (external)	Bays	Balconies	Porches	Cornices etc.	Copings	Parapets	Renderings	Cladding	Panel elements	Loadbearing elements
Carbon steel	METALS	○	●	●	○	●	●		○	○	○	●	●	●	●	●	●	●		○	○					○	●	○
Galvanised steel	METALS	●	●	●	●			●	○	●	●								○				○			●	●	
Galvanised/bare steel	METALS											●		●														●
Wrought iron	METALS		●												●			●		●	○							
Cast iron	METALS				●				●									●		●								○
Stainless steel	METALS	○	○	○	○	○	○	●	○	○	○		●	○	○	●	○	○	○	○	○		○	○		○	○	○
Aluminium sheet	METALS								●	●	●								●		●		●			●	●	
Aluminium alloys	METALS		○	○	●	●		○	○	○	○		●		●	●	●	○	●		●							
Copper	METALS					○		○	●	●	●																	
Bronze, brass etc.	METALS	○	○		○	○	○		○	○	○			○	○	○	○	○	○	○	○					○	○	
Zinc	METALS				●	○	○	○	●	●	●		●		○	○	○			○	○		●			●		
Lead	METALS			○		○	○	○	●	●	●								○	○	○					●		
Limestone	NATURAL STONE	●	●			●		●				●	●		○	○	○		●	●	●	●	●	●		●	●	●
Sandstone	NATURAL STONE	●	●			●		●				●	●		○	○	○	●	●	●	●	●	●	●		●	●	●
Marble	NATURAL STONE												●		○	○	○	●	●	○			●			●	●	
Slate	NATURAL STONE	●	●			●					●		●		○	○	○	●	○	●	●		●			●		
Clay bricks	BRICKS	●	●	●		●		●				●	●		○	○	○	●	●	●	●	●	●	●		●		●
Flintlime bricks	BRICKS																						●	●		●		●
Sandlime bricks	BRICKS																						●	●		●		●
Concrete bricks	BRICKS	●	●					●										○		○		●	●	●		●		
Portland cement	CEMENT AND CONCRETE	●	●	●				●	●				●		○	○	○	●	●	●	○	●	●	●	●	●	●	●
High alumina cement	CEMENT AND CONCRETE																			●				●				●
Reinforced concrete	CEMENT AND CONCRETE	●	●	●				●				●	●		○	○	○	●	●	●			●	●			●	●
Asbestos cement	CEMENT AND CONCRETE	●	●	●	●		○	○	○	●	●		○		○	○	○									●	●	
Lime cements	CEMENT AND CONCRETE							○	○				○		○	○	○				○				●			○
Clay tiles	TILES	●				●					●		●		○	○	○	●			●		●	○		●		
Concrete tiles	TILES	●				●					●												●			●		
GRF Plastic	PLASTICS			○	○		●			●	●				●	●	●		○							●	●	
PVC Sheet	PLASTICS						●			●	●															●	●	
Gypsum plaster	PLASTER																					●			●			
External renderings	PLASTER	●	●			●		●	●			○	○		○	○	○	●	○	○	●	●		●	●			○
Stucco	PLASTER														○	○	○					●		●	●			○
Oil-based paint	PAINT	●	●	●	●	●	●		●			○	●	●	●	●	●	●	●		●	●				●	●	○
Cement paint	PAINT	●	●	●		●		●	●			●	●		○	○	○	○	●	○	●	●		●	●	●	●	
Thick paint	PAINT	●	●	●		●		●	●			●	●		○	○	○	●	●	○	●			●	●			

● PRIMARY CONSTITUENT ○ SECONDARY COMPONENT

Figure 3.1 Matrix cross-tabulation of building materials vulnerable to atmospheric sulphur attack, and relevant elements in buildings.

probably bacteria) also affect the weathering characteristics of stone and other porous masonry materials. It is important to take all these into account when assessing actual or potential pollutant damage.

In the main, it is only the external envelopes of buildings that are at risk. However, pollutants generated outside a building can affect materials inside, for example leather, wood or stone antiquities in museums. The levels of externally generated pollutants inside buildings are much lower than outside, but are still at levels at which damage to fragile objects occurs although the levels inside are not likely to affect the building structures *per se*.

3.3 WHICH PARTS OF BUILDINGS ARE MOST LIKELY TO BE AFFECTED?

The location of a particular material in the building structure will have a considerable effect on the damage it may suffer. The most obvious example being the internal features of a building which will undergo very little damage if the structure remains waterproof.

The effect of external location relative to the durability of natural building stones was included in Leary (1983) (Figure 3.1). The external features of buildings can be 'zoned' according to the weathering stone in those areas will undergo, Zone 1 being the most and, Zone 4 the least affected. Leary (1983) recommends limestones of differing durabilities that are suitable for each zone. However, it must be remembered that weathering due to 'natural' forces will occur in exposed positions and that damage resulting from acid deposition (eg sulphate crusts) or salt action can also occur in sheltered positions. Similar studies of other materials do not appear to have been undertaken.

Problems and damage to materials can also be exacerbated by the relative positions of different materials (Schaffer 1932). For example two metals in contact, the expansion of corroding iron in concrete, the staining of stone by metal cladding, or the presence of limestone or lime mortar adjacent to sandstone or granite. In many cases these problems can be overcome by careful selection of materials at the design stage but are still evident on historic buildings and monuments.

3.4 WHICH TYPE OF BULDINGS ARE AT RISK?

The most obvious types of building at risk are those built of natural stone and of these historical buildings (eg cathedrals, churches, castles, historic houses) (Figure 3.2) have received most attention. There is no real problem in establishing where such buildings are and what they are made of. However, the situation may be complicated by the use of materials from different sources and of different durability over hundreds of years.

Modern buildings are stucturally different from the older stone cathedrals and churches or timber frame buildings and have been exposed for considerably less time. Consequently they are less at risk. Smaller modern buildings are mainly of brick or block, or frame and panel construction. The larger ones are of frame construction, for example steel or concrete, with some type of cladding (concrete, stone, metal or plastic) on the outside (Figure 3.3). Modern buildings also have the advantage that they can be designed, constructed and protected to avoid environmental damage and degradation wheras an historic building cannot.

One other class of buildings that has received study is agricultural buildings that are constructed mainly of steel (often galvanised sheet steel). Rates of corrosion and replacement of these buildings have been compiled by Shaw (1978) and MAFF (1986), and further analysed by Saunders (1983).

3.5 WHAT IS AT RISK? – ESTIMATING THE AMOUNT OF MATERIAL AT RISK

It is difficult to establish a comprehensive inventory of buildings, their constituent materials and how much they are at risk. Records of designs and specifications often exist, especially for modern public buildings, but the data cannot always be located or made available.

In order to achieve a reliable estimate of the damage to building materials from acid deposition, it is necessary to estimate or calculate the total quantities of materials at risk, ie the total quantity of exposed surface area of those materials known to be susceptible to damage by acid deposition. In addition, it is desirable to know the degree of exposure of these materials to the damaging agents in the environment.

Essentially, there are two basic approaches that can be used to determine the quantity of materials at risk:

(i) The Inventory/Census - whereby each building is individually examined (building-by-building inventory).

(ii) The Probability Distribution Approach.

Limestone durability class	Inland				Exposed coastal			
	Low pollution		High pollution		Low pollution		High pollution	
	No frost	Frost	No frost	Frost	No frost	Frost	No frost	Frost
	Suitability zones for various limestones in a range of climatic conditions							
A	Zones 1–4	Zones 1–4	Zones 1–4	Zones 1–4	Zones 1–4	Zones 1–4	Zones 1–4	Zones 1–4
B	Zones 2–4	Zones 2–4	Zones 2–4	Zones 2–4	Zones 2–4	Zones 2–4	Zones 2*–4	Zones 2*–4
C	Zones 2–4	Zones 2–4	Zones 3–4	Zones 3–4	Zones 3*–4	Zone 4	—	—
D	Zones 3–4	Zone 4	Zones 3–4	Zone 4	—	—	—	—
E	Zone 4	Zone 4	Zone 4*	—	—	—	—	—
F	Zone 4	Zone 4	—	—	—	—	—	—

* Probably limited to 50 years' life

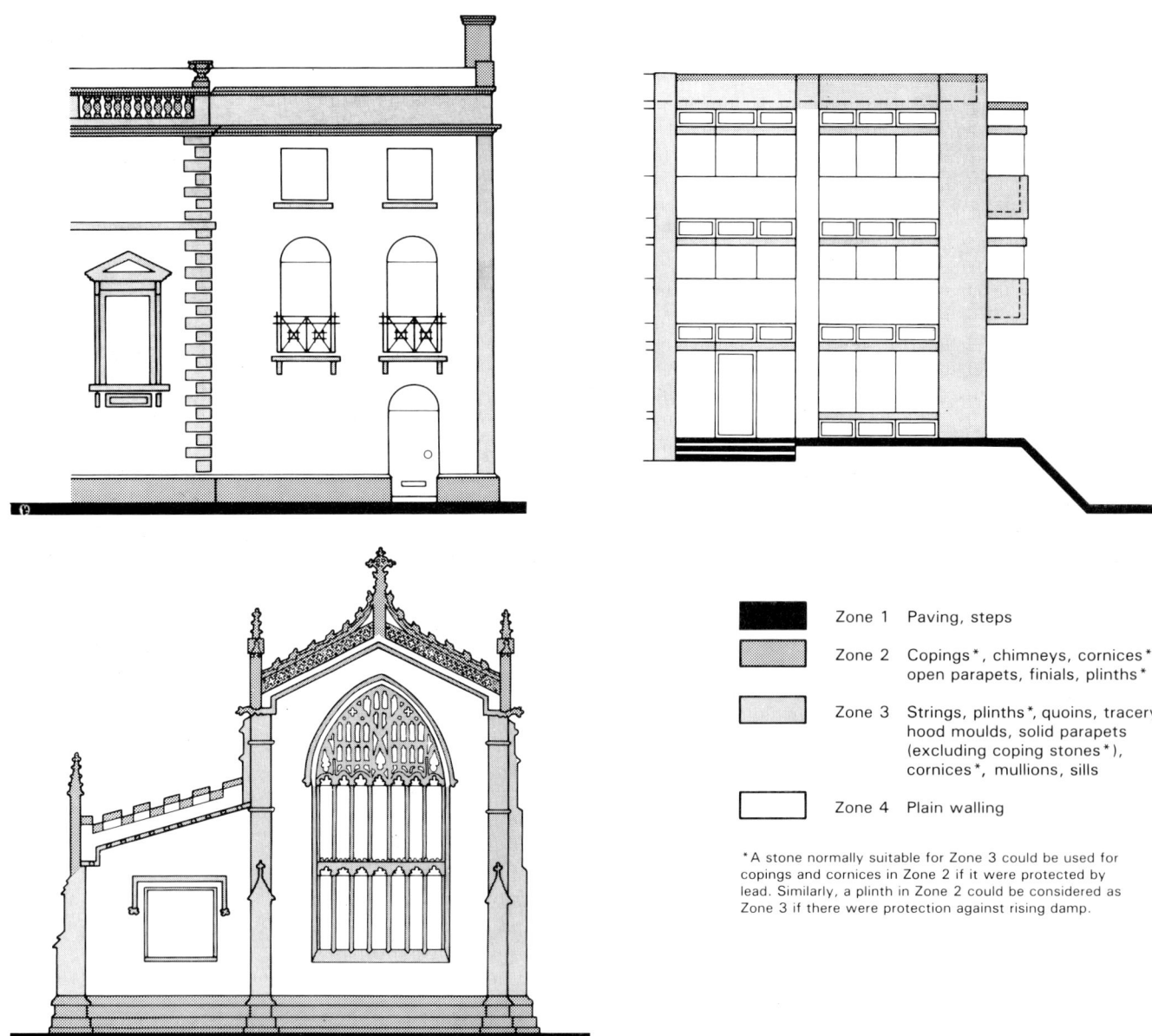

Zone 1   Paving, steps

Zone 2   Copings*, chimneys, cornices* open parapets, finials, plinths*

Zone 3   Strings, plinths*, quoins, tracery hood moulds, solid parapets (excluding coping stones*), cornices*, mullions, sills

Zone 4   Plain walling

*A stone normally suitable for Zone 3 could be used for copings and cornices in Zone 2 if it were protected by lead. Similarly, a plinth in Zone 2 could be considered as Zone 3 if there were protection against rising damp.

Figure 3.2 Effect of environment on the suitability of limestones for four exposure zones on a building.

3.5.1 The Inventory/Census Method

This method is most successful when applied to 'cultural' buildings which are essentially unique and the number of buildings is relatively limited (cathedrals, churches, etc). The numbers of other important buildings (castles, historic houses, listed buildings) are more difficult to establish because of their greater numbers and the absence of complete lists. Further difficulties are encountered with groups of buildings forming a single complex, conservation area and with artistic features such as statues. Also, it is not enough to know simply how many 'historic' or 'cultural' buildings exist, without details of their materials and design.

Lists of buildings are available through the National Trust (eg National Trust 1983), Property Services Agency (1982; 1983; 1984), the Royal Commission on Historic Monuments (RCHM, county by county coverage), through local government lists of graded buildings, and the Church Commissioners. The effort required to construct a database from these can be judged by the RCHM's having spent over 70 years on a similar task without reaching completion.

3.5 The Probability Approach

In this method, all areas of exposed materials are aggregated to provide a probability function of the exposed material per unit area of land. This method is best suited for buildings which have the characteristic that many of them are practically indistinguishable from each other.

However, the probability distribution methods used to date in economic assessments and cost estimate studies appear to be poor at characterising and generalising. For example, the OECD (1981) review used national data on production and consumption figures (in terms of total tonnage per annum) for galvanised sheet and profile steel. As assessment was then made as to the proportion of material actually exposed and at risk to acid deposition and air pollution in general. These data were then related to estimates of acid deposition at 20 km square intervals, a method that precludes comparisons being drawn at local levels.

A similar approach, based on production figures, has also been followed in the past by the CEGB for the calculation of zinc or zinc-coated steels and wire.

Both the OECD and CEGB approaches suffer from a limitation in that they are applicable only to materials for which production data exist. In addition, a series of assumptions has to be made in order to calculate the amount of material actually in service (or in place) and exposed to deposition. These assumptions include:

— the percentage of total tonnage used for building purposes

— the percentage of material externally exposed

— the percentage of material withdrawn from service (end of service life)

The same basic premise is employed by Salmon (1970) except that in this study, national estimates of materials in place were derived from national sales (eg dollars not tonnage).

McFadden and Koontz (1980) used fire insurance maps to estimate the distribution of primary wall and roof materials only. The shortcomings of this approach are obvious in that it aggregates together all kinds of exterior masonry and only certain materials are included. Other materials such as flashings, gutters and fences are ignored and this leads to a consistent under-reporting of exposed materials. In addition, the data were site specific to two particular cities in the USA.

TRC Environmental Consultants (1983) developed a systematic ground survey method, based on random samples, for a pilot project in Boston, USA. A 100 × 100 ft grid was defined and randomly selected locations were investigated by compiling an inventory of all materials contained within the imaginary 100 ft square boundaries. Unfortunately this procedure resulted in fractions of buildings and vacant spaces being included in the inventory. It also ignored the relative amounts of different types of land use and their spatial distribution and when extrapolated for the entire city of Boston the results were inaccurate.

The US Geological Survey (USGS), National Mapping Division (Rosenfield 1984b) and the US Army Corps of Engineers, Cold Regions Research and Engineering Laboratory (CRREL) (Merry & LaPotin 1985) developed a stratified random sampling scheme, based on six strata or sampling frames:

(i) Urban Centre, Business District

(ii) Urban Livelihood, Industrial, and Commercial

(iii) Urban Multi-family Residential

(iv) Urban Single-family Residential

(v) Suburban

(vi) Rural

The statistical design of the sample was based on a multinomial distribution, assuming that most buildings would contain no more than five different materials types. A target precision of ± 20% required of about 100 randomly selected sample points in each frame. It was realised that sampling errors would be considerably larger for those minor materials occurring only as small fractions of a complete building. The concept of a sample 'footprint' or quadrant was adopted in order to ensure sampling of equal land areas at each location within a given sample frame. In order to maintain a roughly constant probability of finding a building within the sample area, the quadrant size was adjusted according to the average population density.

Other, less well defined, methods have also been promoted in the past; these include the Spatial Sample Design (Rosenfield 1984b). This method assumed five construction materials to be of interest in building walls and roofs. The sampling frame consists of a number of 'census tracts' having a certain commonality of population density, single unity dwellings, and mix of land use. Each building is considered to be a cluster of elements composed of single construction materials.

The European Commission (CEC) and the United Kingdom Department of the Environment (DOE) have been supporting a programme of work on both historic and modern buildings (Williams 1985). This has, amongst other objectives, the development of improved stock-at-risk inventory methodologies and the development of models representing building damage mechanisms. The work on building damage mechanisms is concerned with deriving a base line of current knowledge on the various causes of building damage and assessing the relative importance of air pollutants to this damage. Such data will then be used in interpeting estimates of damage cost.

The CEC/DOE study has concentrated on developing a 'stock-at-risk' methodology which allows an estimate of the total quantity (surface area) of exposed materials in buildings to be obtained and also re-defines 'stock-at-risk' in terms of building components. The methodology involves deriving 'indentikits' for different types of buildings. The building stock is categorised into 15 categories of building ranging from single family units to industrial premises. Each building type is specified by the following: number of stories, floor space index, basal surface index, number of buildings per square kilometre and age of building.

It is possible to estimate the area of the external envelope (wall, window, roof) using standard quantity surveying techniques and field measurements for each building category. Field surveys are then used to derive the typical materials used in the construction of the external envelope (galvanised steel, concrete, brick, cement roof tiles, etc). Land-use maps which show the different building type patterns, are then used to determine the proportion of each km grid square occupied by a particular building type. This methodology has been applied in four cities: Birmingham and Lincoln in the United Kingdom, Dortmund and Cologne in West Germany.

The DOE/CEC study also includes an assessment of the condition of exposed materials. Both the substrate materials and the surface finish are classified into one of three classes: unaffected, slightly affected and badly affected by atmospheric weathering. A method developed from this study gives a series of settlement patterns based on a comprehensive list of building categories. These patterns are recognisable from 1:10,000 maps. This project has also re-defined the 'stock-at-risk' function of buildings and building components (eg gutterings, window units, window sills, etc). An inventory of building components using 'identikits' has also been obtained for Birmingham, Lincoln, Dortmund and Cologne (Williams 1985).

In the Nordic countries, a joint programme on assessing the effect of reduced SO_2 emission on corrosion damage is currently being carried out. The Swedish and Norwegian projects are dealing mainly with the question of the 'stock-at-risk' inventory, while in Finland the work is concentrating on dose-response relationships. The Swedish and Norwegian inventories are being carried out using identical sampling methods and building survey protocols. The Swedish project is covering the materials at risk in Stockholm, mostly affected by many local sources and the Norwegian survey is concentrating on smaller industrial towns (eg Sorpborg) affected by large industrial sources of SO_2.

The sampling method and building survey protocols are based partly on earlier studies by the Swedish Institute for Building Research which examined the condition of the Swedish building stock. The sampling procedure is based upon classifying communities into four levels based on annual, mean, and winter SO_2 concentrations. Buildings are grouped into 9 categories: single family houses pre-1920, 1921-1960, 1961-1984; apartments pre-1920, 1921-1960, 1961-1984; farm houses, industry, and shops and offices. Data on real estate taxation is used to identify target sample buildings which are then surveyed. A total of 400 buildings will be surveyed. The building survey includes obtaining information on the local environment, design data, a materials inventory, and a description of the condition of the building. The materials inventory is sub-divided into 9 building components: foundations, walls, windows, doors, balconies, roofs, guttering, ancillary buildings (eg garage), and fences. Once time studies have been performed they will form the basis of a model to be be used in other Nordic cities.

3.5.3 Conclusions

Estimating the number of buildings and the amount of material at risk from acid deposition is clearly a difficult and complex process. However, it is essential for any estimate of the cost of damage and consequently for any form of cost-benefit analysis. The two main alternative methods each have their advantages and disadvantages. Using the Inventory/Census method all buildings must be recorded in detail but accurate results could in principle be achieved given sufficient time and resources.

The Probability-Distribution approach appears to be the more successful and more suitable for modern buildings. The following specific conclusions can be arrived at:-

(i) It is possible to use an approach based upon building type models to produce an estimate of the stock of materials and components at risk and then utilise this for evaluating the costs of damage.

(ii) The costs of damage to building materials by atmospheric pollution will depend not only on the material type used, but also on the nature of the component and its function within the building.

(iii) Using building components as the central focus for assessing damage and damage costs allows aspects of service life and maintenance cycles to be incorporated into the damage cost assessment process.

(iv) Given that the methodology relies upon the use of maps, known costs and service lives, etc it can be used at either a regional or national scale. There is a need to calibrate certain of the building category 'identikits' (especially housing and vernacular architect use) to the specific materials used in different regions.

(v) The cost assessments made are applicable to all buildings, including the general stock of historic buildings in conservation areas and as such produce a minimum estimate of the willingness to pay to reduce damage.

(vi) Historic buildings and monuments of national and international importance require a quite different evaluation approach since their cultural and aesthetic significance may be much greater than their cost of replacement. Hence the inventory/census method is more appropriate for this category.

4 Trends in air pollution emissions and air quality in the United Kingdom

---

SUMMARY

Annual average concentrations of SO_2 in urban atmospheres in the United Kingdom have decreased significantly in the last 30 years, reflecting the proportional shift in emissions from low and medium level urban sources to high level emitters often located outside urban areas.

Annual average urban concentrations of black smoke have decreased by a factor of ten during the last 30 years.

Over the past 50 years emissions of NO_x in the United Kingdom have roughly doubled, primarily because of the increase in road traffic. However, there has been no apparent trend in urban air concentrations of NO_x or

ozone at the relatively few sites where measurements have been made over the past ten years or so.

Elevated concentrations of CO_2 in urban areas may affect rainfall pH but further study is required.

Chloride concentrations in rainwater are generally dominated by marine sources. In industrial areas, however, other sources will be important. Emissions and ambient air chloride concentrations arising from coal combustion are likely to have decreased markedly over the past 30 years or so, particularly in urban areas.

4.1 INTRODUCTION

The evidence in Chapter 2 of this report shows that buildings are still being damaged by air pollution although there have been significant changes in pollutant emissions and air quality in the United Kingdom since the turn of the century and in particular since the 1950s and 60s. These trends are important because they are central to the debate on acid deposition effects on buildings and in particular to the question of whether or not changes in pollutant emissions and air quality have been accompanied by changes in rates of damage.

4.2 SULPHUR DIOXIDE

4.2.1 United Kingdom Emissions of Sulphur Dioxide

Sulphur dioxide emissions arising from man's

activities are calculated annually by Warren Spring Laboratory (WSL) and are published in the annual DOE Digest of Environmental Protection and Water Statistics (1978-1986). Emissions have also been calculated for the period 1853 to 1974 by Bettelheim and Littler (1979) and their estimates are compared with those of WSL in Figure 4.1. The two sets of calculations are based on similar figures for fuel use. The differences between the two sets of figures, which amount to between 5 and 10%, arise chiefly from uncertainties over sulphur contents of fuels, different assumptions regarding the amounts of sulphur retained in the ash following coal combustion and the fuel use categories included by the two calculations. The two plots consequently represent an indication of the precision of the estimations. Bearing in mind that the emissions are calculated from fuel deliveries rather than use, for many categories the overall confidence in the recent estimates (since 1950) is probably nearer 15-20% but the earlier data must be considered even more uncertain.

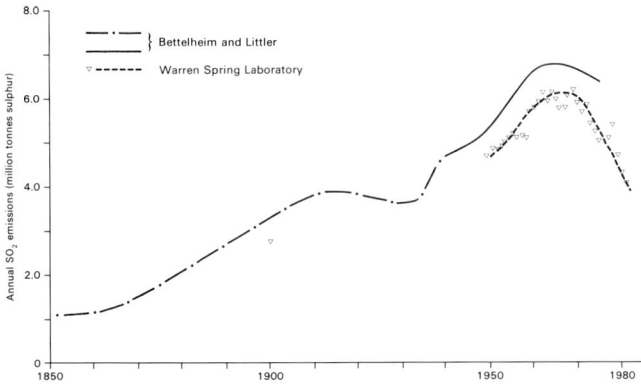

Figure 4.1 Sulphur dioxide emissions in the United Kingdom from 1850.

Both sets of estimates yield similar trends, rising steadily from a total of 1.3 Mt in 1850 to a peak in the decade 1960-1970 followed by a marked decrease in the 1970-80s. More detailed data for individual years since 1965 are shown in Table 4.1 from which it can be seen that from peak levels of 6.1-6.6 Mt in 1970, United Kingdom emissions have decreased to levels similar to those obtaining in 1910.

Table 4.1 United Kingdom SO$_2$ emissions from 1965-1984.

	United Kingdom SO$_2$ Emission (million tonne)	
Year	WSL	Bettelheim and Littler
1965	6.01	
6	5.94	
7	5.71	
8	5.74	
9	6.02	
1970	6.09	6.62
1	5.86	
2	5.65	
3	5.80	
4	5.35	5.94
5	5.13	
6	4.98	
7	4.98	
8	5.02	
9	5.34	
1980	4.67	
1	4.22	
2	4.01	
3	3.69	
4	3.54	

The reduction since 1980 has been due to a number of factors including energy economies (~4%), reductions in sulphur contents of fuels (~4%), changes in fuel use patterns (eg to natural gas) (~5%) and industrial modernisation (~6%).

A graph of source contributions to emissions of SO$_2$ is shown in Figure 4.2. In 1950 low and medium level sources (domestic, commercial, and industrial users) were responsible for approximately 80% of SO$_2$ emissions, while by 1970 this proportion was nearer 50% (Weatherley *et al* 1975). By 1984 the proportional contribution from low and medium level emissions had decreased still further to 25%, with power stations now contributing some 70%. These trends have important consequences for urban (and rural) air quality in terms of both absolute concentrations and the proportional contribution from the major source categories.

4.2.2 Sulphur Dioxide Concentrations in the United Kingdom

Following the London smogs of the early 1950s and the Clean Air Act of 1956 a continuing programme of smoke control was introduced in the United Kingdom. In order to provide comparable data of a consistent minimum quality for the wide range of areas (urban, industrial, rural) found in the United Kingdom, the National Survey of Smoke and Sulphur Dioxide was set up in 1961 by the Department of Scientific and Industrial Research (DSIR) and co-ordinated by Warren Spring Laboratory (WSL). The monitoring sites numbered about 500 initially, increasing to about 1200 by 1966. At present 450 urban sites are in operation.

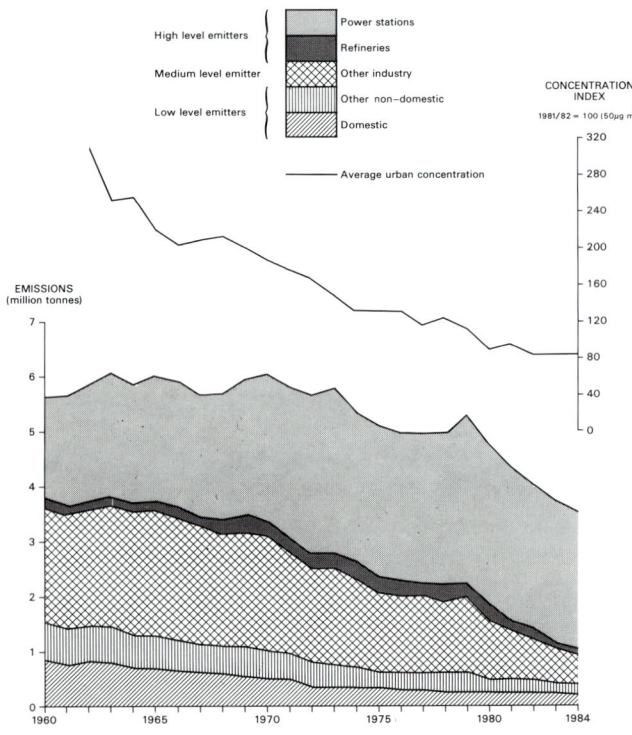

Figure 4.2 Trends in sources of SO$_2$ emissions and air quality.

The reduction in annual average SO$_2$ concentrations averaged over all urban sites in the national survey is shown in Figure 4.2. The decrease in concentrations parallels the decrease in emissions from the low and medium level sources. The relative contribution from different source categories is, over distances of 10 km, strongly dependent on stack height. They will obviously vary from one urban area to another depending on the source mix and can only be quantified by a detailed study of each individual area. Nonetheless over the past thirty years or so, large decreases in urban SO_2 concentrations have clearly arisen, primarily caused by the decrease in domestic and industrial/commercial emissions.

The foregoing discussion presents the national picture, but in order to give some indication of the changes in exposure to SO_2 for buildings and locations of interest, time series plots for individual sites have been analysed. Two of the longest time series for airborne SO_2 concentrations which exist in the United Kingdom are those for County Hall, Westminster (OS Grid reference TQ307797) and for the site at Surrey Street in Sheffield (SK354871), (Identified as National Survey Site: Sheffield 2). Virtually complete records exist for both sites from 1932/3. Results from the London sites, obtained using peroxide bubblers, are shown in Fixture 4.3. SO_2 concentrations here and in Sheffield remained broadly constant at 300-400 μg m^{-3} up to the early 1960s and thereafter fell rapidly.

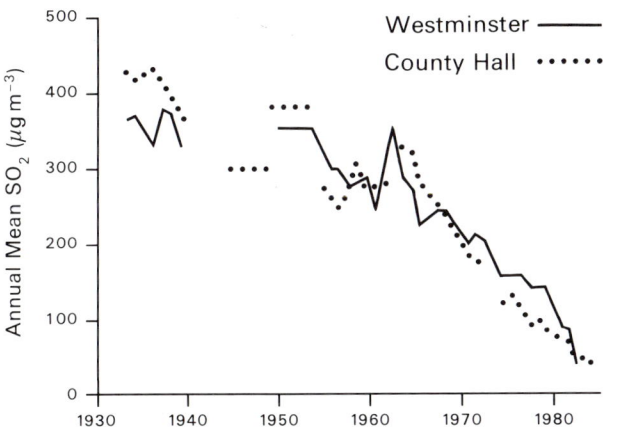

Figure 4.3 Trends in SO_2 concentrations in Westminster, London over the last 50 years.

Similar time series have been plotted for sites close to buildings of interest, although of necessity the time series are shorter than those shown in Figure 4.3. Data for the two sites in the National Survey closest to St Paul's Cathedral, London are shown in Figure 4.4. The 98th percentile of daily average values over a year are plotted in addition to annual means. The 98th percentile concentration in this context is that concentration exceeded on 2% of days (7) in a year.

Concentrations in other urban areas of the United Kingdom have decreased in a similar manner. At urban sites in York and Lincoln, where there is concern over damage to historic buildings, annual SO_2 concentrations have decreased from 150-170 μg m^{-3} in the early 1960s to current levels of 40-50 μg m^{-3}. Peak daily mean concentrations as measured by the 98th percentile over a year have also decreased during this period from 400-650 μg m^{-3} to 100-120 μg m^{-3}.

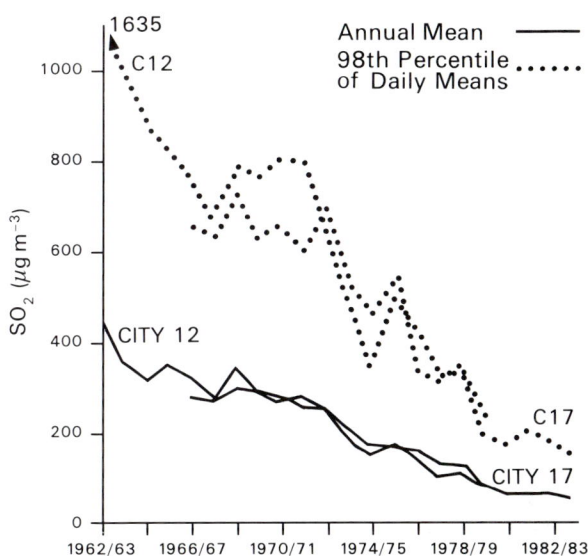

Figure 4.4 Trends in SO_2 concentrations in the City of London between 1962/3 and 1983.

Maximum daily SO_2 concentrations were reached during 'smog' episodes. During December 1952 daily average concentrations of 3,000-4,000 μg m^{-3} were measured at some sites in London, and in December 1962 some sites measured concentrations greater than 4,000 μg m^{-3} on two days. These figures given an indication of the peak/mean ratios for the higher percentiles. A more detailed report on a high pollution period in December 1975, when daily mean SO_2 concentrations of up to 1,200 μg m^{-3} were observed at some sites in London, is given in Apling *et al* (1977).

Information on urban SO_2 concentrations before the 1930s is scarce. An analysis of coal imports into London from 1600 AD onwards (Brimblecombe 1977) suggests that coal use per unit area was broadly constant from 1800-1900, but that there was a marked decrease (by ~50%) in SO_2 concentrations from 1900-1950. The reasons for this may lie in the model used. The measured data suggest that SO_2 annual average concentrations stayed broadly constant from 1930-1960, a period of increasing oil and decreasing coal consumption. Nonetheless it seems likely that SO_2 concentrations in London during the 19th century were at least as high as those measured in the 1930s, ie 300-400 μg m^{-3} annual average. Using a ratio of 2.8 as a typical value of

23

98th percentile/annual mean ratio for SO_2 (obtained from an analysis of National Survey data) suggests that 98th percentiles of 850-1200 $\mu g\ m^{-3}$ were likely in London during the 19th century. Peak values in years with more extreme winters could have been even higher.

4.3 SMOKE

Since 1960, detailed estimates of smoke emissions have been calculated from coal combustion records, these are shown in Figure 4.5. It is clear that there has been a major decrease in overall emissions of smoke from coal combustion, especially in the domestic sector as areas under smoke control have increased. This decrease in emissions has had a marked effect on urban smoke concentrations (Figure 4.5). Annual mean urban smoke concentrations in the early 1960s were, on average, $\sim 140\ \mu g\ m^{-3}$, with some areas reporting concentrations of up to 350 $\mu_{\circlearrowleft}\ m^{-3}$. Estimates for historical smoke concentrations have been made by Brimblecombe (1977). However, this report quoted concentrations averaged over the whole of London, so his values may not be strictly comparable with values quoted above. His figures suggest that concentrations in London peaked in 1900 and declined by about 50% up to 1930-40. This suggests that smoke concentrations in London during the 19th century were as high as those in the mid 20th century and could well have been up to a factor of two higher in the late 19th century.

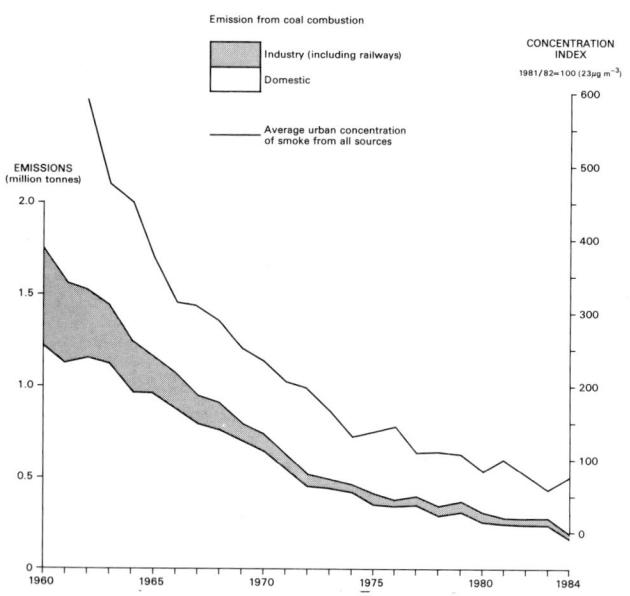

Figure 4.5 Trends in sources of smoke emission and air quality.

In the late 1950s to early 1960s, 98th percentile smoke concentrations were typically 550-1300 $\mu g\ m^{-3}$ (~ 3.8 times annual mean). Peak

daily concentrations of smoke reached 4000 $\mu g\ m^{-3}$ at two sites in London in December 1962. It is estimated that, averaged over London, concentrations could have been up to twice as high in the smog episode of December 1952 as in the 1962 episode.

Smoke concentrations are now very much lower and the United Kingdom average of urban annual means is now approximately 17 $\mu g\ m^{-3}$. Smoke annual mean concentrations in Central London are typically 25-35 $\mu g\ m^{-3}$, while those in York and Lincoln are 15-20 $\mu g\ m^{-3}$ and 15 $\mu g\ m^{-3}$ respectively.

4.4 OXIDES OF NITROGEN

Estimates of United Kingdom emissions of NO_x from 1905 to 1980 have been made by WSL and are shown in Figure 4.6 disaggregated by fuel type. The broad trend is one of roughly constant emissions until about 1945. Since then emissions increased by a factor of approximately two. There is no indication that NO_x emissions peaked in the late 1960s/early 1970s. Recently however WSL have improved the method of calculating motor vehicle emissions (Rogers 1984) and the estimates of NO_x emissions since 1983 have been revised

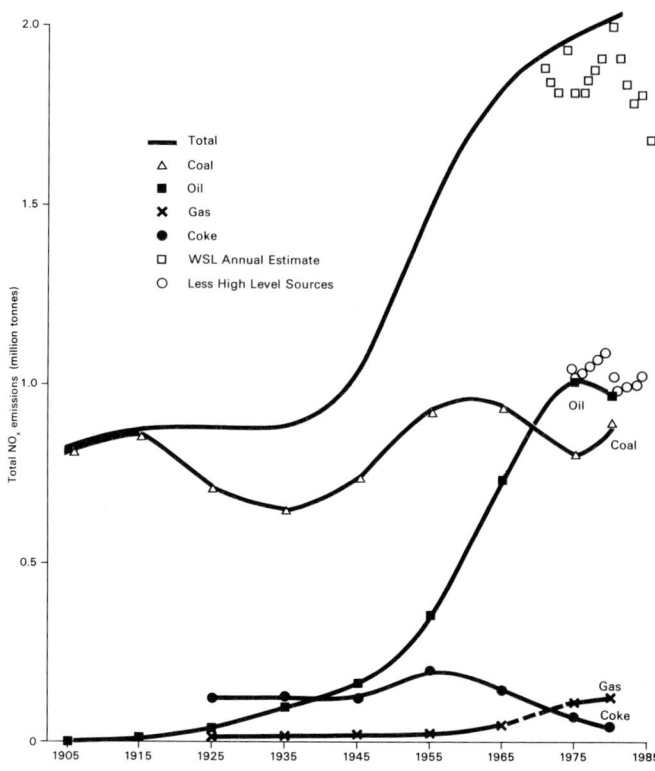

Figure 4.6 NO_x emissions in the United Kingdom since 1905.

24

upwards by 10%. Revision of earlier years is difficult to carry out accurately as the WSL measurements underpinning the new method relate to the United Kingdom car fleet in 1981-1983.

However, approximate revisions for earlier years have been carried out and these are shown in Figure 4.6. The data since 1974 show no clear trend in national NO_x emissions, either in total NO_x or in the emissions from low and medium level sources (ie excluding power station and refinery emissions). There is an indication that total NO_x emissions decreased between 1979 and 1984, largely due to a decrease in this period of some 250 kt year^{-1} in power station emissions. 1984 was an anomalous year because of the dispute in the coal industry. It is clear that since about 1945 there has been a significant increase in United Kingdom NO_x emissions, chiefly due to increased oil consumption by the transport sector.

Air quality data for nitrogen oxides in the United Kingdom are very much less plentiful than is the case for SO_2. The longest run of measurements at an urban location is 11 years at the WSL Central London Laboratory in Victoria. These data show no significant trend in NO_x concentrations over the run of data. Peak hourly concentrations are determined more by variations in meteorological conditions than by emissions and tend to be a less reliable indicator of trends than are longer term averages. Typical values of annual mean NO and NO_2 concentrations in Central London are ~40-65 μg m^{-3} and ~60-80 μg m^{-3} respectively. Peak hourly concentrations, such as 98th percentiles over an annual period, can be 3-4 times the annual mean for NO and 2-3 times for NO_2.

4.5 CARBON DIOXIDE

Carbon dioxide is not usually considered to be a pollutant gas. However as Chapter 2 and 6 show it does add to the acidity of rainwater and cause some degradation to limestone and concrete. The burning of fossil fuels has caused an increase in atmospheric concentrations of CO_2 from approximately 290 ppm in 1870 to 340 ppm in 1985 (Brimblecombe 1986; Clarke 1969; Smith 1975). It is predicted that the concentration could be 600 ppm in 2030-2050 (Brimblecombe 1986). Amorosa and Fassina (1983) suggest that peak urban concentrations can reach 3000 ppm and that the pH of rainfall in such episodes shows a decrease by 0.5 units. As a consequence, attacks by CO_2 on calcareous stone proceeds more rapidly because the CO_2 concentration increase leads to a higher partial pressure and more decay (Winkler 1970) and directly because of the lower pH.

However, studies described in Chapter 6.4.3 (Guidobaldi 1981; Guidobaldi and Mecci 1985) claim that the rate of calcium carbonate dissolution was insensitive to pH above 4.0. It is unclear if peaks episodes of CO_2 created and confined within street canyons could effect rainwater in the same way.

4.6 OZONE

Measurements of ozone concentrations over extended periods are scarce in the United Kingdom. Unlike the pollutants considered so far, ozone is not a primary pollutant. Ozone is produced in high concentrations in the stratosphere by UV irradiation and transported into the free troposphere to supplement ozone produced there photochemically. The 'background' tropospheric concentrations can be enhanced in the atmospheric boundary layer (the lowest 1-2 km or so of the atmosphere) by ozone produced in so-called 'photochemical episodes'. These involve the action of sunlight on ozone-generating precursor pollutant emissions of nitrogen oxides and reactive volatile hydrocarbon/organic compounds. For a more detailed discussion of atmospheric ozone, the reader is referred to a recent review (Derwent 1986) and the report of the Photochemical Oxidants Review Group (PORG 1987) which was set up, by the Department of the Environment, initially to assess air quality data in the United Kingdom.

Measurements of ozone in the United Kingdom have been mainly at rural sites because urban concentrations tend to be reduced by primary emissions of NO. Annual average concentrations in urban areas are in the range 20-40 μg m^{-3} with rural concentrations in the range 40-60 μg m^{-3}. 98th percentiles of hourly averages in a year are generally 70-150 μg m^{-3} in urban areas and 100-175 μg m^{-3} in rural areas. The longest set of ozone measurements in an urban area in the United Kingdom is that obtained by WSL at the Central London Laboratory. It is not possible to discern whether or not there is a long-term trend because of year to year variations (PORG 1987). The comments made above on the predominant role of short term meteorological variations in determining short period (eg hourly) concentrations of NO_x are even more relevant for the secondary pollutant ozone. In this case production of elevated ozone concentrations not only depends on poor dispersion conditions but also sufficient sunlight has to be present together with favourable air mass trajectories transporting precursor pollutants from source to receptor.

4.7 CHLORIDE

There has been no definitive calculation of chloride (Cl^-) emissions in the United Kingdom but some estimates of the major sources have been made.

An average Cl content of 0.25% is appropriate for coal in the United Kingdom. Assuming 95% of the chlorine is emitted to atmosphere this implies an HCl emission of 240 kt year^{-1} for 1983 from United Kingdom coal consumption. The use of coal with a higher Cl content (0.6%) in some areas may lead to higher concentrations locally. Emissions from oil fuels are very low by comparison and an estimate of emissions from this source suggests a figure of 1.5 kt year^{-1} for 1983. The incineration of refuse can also be a source of HCl and estimates of emissions based on WSL measurements at several municipal incinerators suggest a national emission of HCl from this source of 5-7.5 kt year^{-1}. Another source of HCl is likely to be that from the manufacture of chlorine and HCl. Very tentative preliminary estimates of emissions from this source suggests an annual emission of 15 kt year^{-1}.

A more detailed assessment of HCl emissions from these sources would be necessary before firmer conclusions could be drawn but nonetheless it is likely that coal combustion represents the major source of national emissions of HCl.

It should also be mentioned that national man-made emissions of Cl^- are broadly an order of magnitude less than SO_2 or NO_x. Furthermore, although detailed data are not available at present, the average Cl content of coal in use in the United Kingdom has only increased by a small amount over the past 30 years, so that ambient concentrations of Cl^- from coal combustion will probably have shown a similarly large decrease to those of SO_2 since the 1950s and 60s.

Due to its maritime climate, all the rain in the United Kingdom contains sufficient Cl^- derived from sea-salt, to cause appreciable corrosion of iron, copper and zinc. This tends to impose a lower limit on the rate of corrosion irrespective of pollutant levels. The concentrations of Cl^- in rainwater tend to be dominated by marine-derived Cl^- and year-to-year variations are generally reflections of meteorological differences rather than trends in pollutant emissions. Annual precipitation weighted mean Cl^- concentrations in rainfall at Goonhilly near the Cornish coast, for example, from 1981-1985 were 290-470 μ eq litre^{-1} whereas the corresponding non-marine chloride concentrations were 24-42 μ eq litre^{-1}. At inland industrial sites with heavy concentrations of coal-burning equipment, rainwater chloride concentrations could however be 1000 μ eq l^{-1} (BISRA 1938). The use of sodium chloride on road surfaces during cold weather may also be a

significant source of chloride for adjacent building materials.

There have been no systematic measurements of particulate Cl^- or gaseous HCl over an extended period in the ambient atmosphere in the United Kingdom. However, data from an 18-month survey at Leeds (Willison et al 1985) indicated almost equal contributions of about 1 μg m^{-3} annual average from marine and non-marine sources to the total particulate Cl^-.

4.8 CONCLUSIONS

(i) Annual average concentrations of SO_2 in urban atmospheres in the United Kingdom have decreased significantly in the last 30 years. Although historical data before the early twentieth century are scarce, it is likely that prior to the 1950s, concentrations of SO_2 in Central London had changed little since the 17th century. Since the 1950's and 60's, urban concentrations of SO_2 have decreased at a faster rate than total emissions of SO_2. This reflects a proportional shift in emissions from low and medium level urban sources to high level emitters often located outside urban areas.

(ii) Annual average urban concentrations of black smoke have decreased by a factor of ten during the last 30 years. An important development has been the change in the relative contributions from the major sources, most notably the shift from domestic solid fuel to vehicle emissions.

(iii) Over the past 50 years, emissions of NO_x in the United Kingdom have roughly doubled primarily because of the increase in road traffic. However, there has been no *apparent trend* in urban air concentrations of NO_x or ozone at the relatively few sites where measurements have been made over the past ten years or so.

(iv) Elevated concentrations of CO_2 have been measured in urban areas and may affect rainfall pH but further studies are required.

(v) Chloride concentrations in rainwater are generally dominated by marine sources. In industrial areas, however, other sources will be important. No definite calculations for Cl^- emissions exist for the United Kingdom but coal combustion appears to be the most important man-made source. Emissions and ambient Cl^- arising from coal combustion are likely to have decreased markedly over the past 30 years or so, particularly in urban areas.

5 Atmospheric modelling and source attribution

SUMMARY

The atmospheric dispersion models discussed in this chapter provide reasonably accurate estimates of the contributions to airborne pollutant concentrations from the major source categories.

In an investigation of a sample of 26 towns in the United Kingdom, the external contribution to 1982 annual mean SO_2 concentrations was estimated to be at least 40% and in most cases 50% or greater. A large proportion of the external contribution arises from United Kingdom power stations.

For episodic conditions in urban areas, calculations demonstrate that low and medium level sources tend to be more important than high level sources.

5.1 INTRODUCTION

The objective of this chapter of the report is to determine the contributions of major source categories of pollutants to atmospheric concentrations and deposition in urban areas. This involves an assessment of the results of numerical dispersion modelling studies and measurements of relevant pollutants in urban and rural areas. Details of the models are given in Appendix A.

5.2 MODELS USED IN THIS REPORT

In this report three models have been used to estimate the concentrations of SO_2 arising from different source categories at four locations in the United Kingdom. The scale of the modelling exercise ranges from urban (1-30 km) to medium range transport (100s of kilometres).

Consequently, each 'model' used in this study may in some cases contain an urban or short-range component, together with a component appropriate to the medium range transport of pollutants. The models used were those developed by the Central Electricity Research Laboratories (CERL), British Coal Corporation (BCC), and Warren Spring Laboratory (WSL).

The models have been used to calculate the contribution to the 1983 annual average ground level concentrations of SO_2 at three sites of interest in relation to damage to buildings and materials, namely York, Lincoln, and London. Calculations have also been carried out for a site at Bottesford operated by the CEGB Midlands Region. This site is not in an urban area, but is in a rural location close to power stations and the urban and industrial areas of the East Midlands.

Aside from the specification of the physical and chemical processes in the models, the pollutant emission rates of the various sources are probably the most important inputs and often the most difficult to accurately specify. In this report the source categories reported are local urban area, United Kingdom power stations, the remaining United Kingdom sources, and European emissions. However, detailed emission inventories are essential, particularly on the urban scale where a spatial resolution of 1 km or better is desirable. In the areas chosen for this report such data are not available. The detailed emission inventory produced for London by the GLC (Ball & Ratcliffe 1979) has been used in some models with approximate methods used to calculate 1983 emissions. An updated inventory for 1983 is being prepared but is not available at the present time. Similarly, detailed inventories are under development for York and Lincoln, but these also are not yet available. Consequently very approximate methods based on population data have been used to estimate and disaggregate SO_2 emissions in York and Lincoln *pro rata* with United Kingdom emissions. A figure of 18.4 kg year^{-1} per capita for total emissions was accordingly used. Similarly, while the total United Kingdom emissions of SO_2 are reasonably well known (Digest of Environmental Protection and Water Statistics), approximate methods have been used to disaggregate the emissions spatially. Emissions from power stations are relatively accurately known and have been calculated from data published in the CEGB statistical yearbook and SSEB annual report. Contributions from European sources were calculated using the WSL model, the results being added to the other models in each case.

The applications of the three models have been carried out independently, and with no attempt to use common input data. This procedure was deliberately selected in order to give a realistic impression of the overall confidence which can be placed in the results. Recent work by Derwent (1986) has shown how decision analysis can be used in the study of modelling calculations uncertainties and this could be considered in future work.

5.3 RESULTS OF MODELLING CALCULATIONS FOR SO_2

5.3.1 1983 Annual Average SO_2 Concentrations

The results of the modelling calculations are shown in Table 5.1. Comparison has been made with values for 1983/4 measured in the United Kingdom Monitoring Networks for smoke and SO_2 operated by WSL and by CEGB Midlands Region for the Bottesford site.

The agreement between the measured values and those obtained by all three models is reasonably good and certainly within the range one might expect given the uncertainties discussed above. There is an indication that the models overpredict in London, with the WSL model giving the highest concentration. This is almost certainly due to the inadequate emissions inventory and, in the case of the WSL model, because the calculation was performed for a point at the receptor with the highest measured value (72 μg m^{-3} at the St Marylebone 9 site). Agreement at Lincoln and York is good, while the models appear to overpredict slightly at Bottesford.

The source contributions for Lincoln show a good agreement between the models on the magnitude of the power station contributions of 20-26 μg m^{-3}.

This represents some 40-50% of the measured or calculated values. In proportional terms, the agreement on the contribution of the other source areas is less good but is acceptable given the uncertainties discussed above.

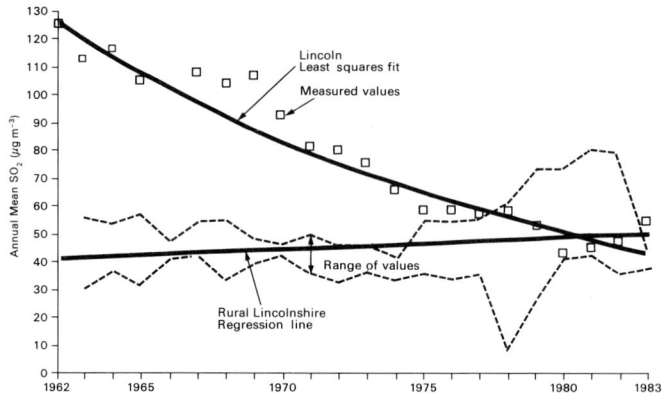

Figure 5.1 Trends in SO_2 concentrations in Lincoln and rural Lincolnshire.

The available measured historical data for the Lincoln urban SO_2 concentrations and data for three sites in rural Lincolnshire (Kirby

Table 5.1 Results of modelling calculations of 1983 annual average SO_2 Concentrations (μg m^{-3}).

Source	York			Lincoln			Bottesford			London		
	CERL	WSL	BCC	CERL	WSL	BCC	CERL	WSL	BCC	CERL	WSL	BCC
Power Stations	16	24	29	20	23	26	20	20	28	10	11-12	19
Local Urban Sources	13	20	8	18	20	8	—	—	—	55	40-51	25
Rest of UK	}11	6	10	}15	6	12	}20	14	10	}39†	6	15
Europe*		3	3		3	3		3	3		3	3
Total	40	53	42	53	52	49	40	37	41	N/A	60-72	62
Observed		37-42			53			28			35-72	

* Sources of SO_2 outside Europe have been ignored; they are likely to contribute at most $\simeq$ 1 μg m^{-3} to annual averages in the United Kingdom.

† This figure includes contributions from the urban area more than 10 km from receptors which are also included in the 'Towns' contribution.

Underwood, Caenby, and Cottam 27) are shown in Figure 5.1. The modelling results are consistent with these historical trends in SO_2 levels. The upper regression line shows the overall decrease in ambient SO_2 concentration in Lincoln and smooths out the scatter of data due to climate variation and measurement errors. The lower regression line (through a great scatter of data points) shows the almost constant historical trend for rural Lincolnshire. Whereas in 1970, concentrations in Lincoln were double those in rural Lincolnshire, concentrations are now very small at urban and rural sites. This suggests that contributions from Lincoln itself are now sufficiently small to be masked by climatic and measurement variability. They may well now be of the order of 10 μg m^{-3} or possibly less. This is borne out by the 1984/5 measured data which shows an annual average of 38 μg m^{-3} at the Lincoln urban site and rural Lincolnshire sites (Cottam sites) in the range 21-39 μg m^{-3}. When taken together these results suggest that the CERL and WSL models may overestimate the contribution from sources within Lincoln. A detailed emission inventory would be invaluable in resolving this issue.

For Bottesford, the CERL and WSL models agree exactly on the magnitude of the power station contribution, while the BCC model calculates a value somewhat higher at 28 μg m^{-3}. These represent a contribution of 47-68% of the total calculated concentrations. All three models overpredict the measured values at Bottesford.

For York the calculated contributions from power stations range from 16 μg m^{-3} in the CERL model to 29 μg m^{-3} in the BCC model, representing some 40-70% of the calculated total concentrations. The urban area of York is calculated to contribute 33-38% of the total. There are, unfortunately, few rural measurement sites near York with which to compare these results. Since the CERL and WSL models appeared to overestimate the urban contribution in Lincoln they may well do likewise for York.

The situation in London is more difficult to model accurately because the large urban area is expected to make a large proportional contribution to ground level concentrations. The absence of an updated emission inventory may also be expected to prove of more significance in London than in smaller urban areas. To some extent the comparison of the modelled and measured results reflects these problems. There is a tendency to overpredict by all three models (the upper end of the range of measured values is that for St Marylebone 9, other sites in this range are central London sites at City 16 (35 μg m^{-3}), City 17 (46 μg m^{-3}) and WSL's Central London Laboratory at Victoria (63 μg m^{-3}). The power station contribution calculated by the CERL and WSL models is 10-12 μg m^{-3}, while the BCC model predicts some 19 μg m^{-3}. The contribution from the urban area is calculated to be 40-70%. In calculating urban contributions the WSL model used the model described by Spanton (1983), with the 1979/80 GLC emissions scaled to represent 1983 values. These estimates suggest that SO_2 emissions in the London area have decreased from 1975/6 to 1983/4 by a factor of just over three. The measured annual average SO_2 concentrations in central London have decreased from values of 150 μg m^{-3} in 1975/6 to concentrations of 35-72 μg m^{-3} in 1983/4. These are consistent with the threefold decrease in an urban contribution which is a significantly large proportion of the total.

As has already been discussed, there has been a major change in the pattern of SO_2 emissions in the United Kingdom since the mid-1970s. The decline in emissions from low and medium level sources, which are chiefly located in urban areas, has resulted in a significant decrease in the contribution to SO_2 concentrations from sources within the towns themselves and a corresponding increase in the relative contribution from external sources. It is worthwhile considering this point in more detail.

However, the lack of adequate emission inventories has already been discussed, so 'local' contributions have been investigated by selecting a sample of United Kingdom towns and cities with a reasonable spread of locations, populations and geographical positions. The average annual mean SO_2 concentrations were calculated for each town and the value of the non-local contribution was calculated from the WSL model. This model has been previously shown to give good agreement with rural SO_2 concentrations, (Perrin (1986) and Appendix A). The 'local' contribution was then estimated by subtracting the non-local concentration from the observed value. When plotted as a function of population the 'local' contributions were in reasonably good agreement with values obtained from the simple urban model of CERL (Appendix A), but there was considerable scatter in the data (Figure 5.2). The simple urban model appears to underpredict the local contribution in areas where there are local sources which are proportionally larger than the national per capita average emission (eg where there is extensive coal combustion). The analysis confirms that the annual mean SO_2 concentration is not always a sensitive function of population and depends more strongly on the local per capita emission rate. Detailed local emission inventories are the only adequate way of obtaining this information.

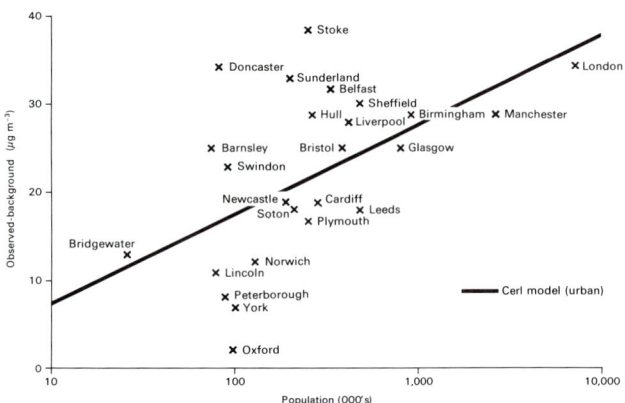

Figure 5.2 Plot of 'local' contributions to annual average SO$_2$ concentrations (μg m^{-3}) against population.

An important conclusion of this analysis is that in the towns studied, the non-local contribution to annual mean SO$_2$ concentrations is estimated to be at least ~40% and in most cases is ~50% or greater. The exceptions to this are those towns with a population of several hundred thousand which lie outside the higher SO$_2$ regions of the United Kingdom (eg those broadly outside the 2 g of sulphur m^{-2} contour or 20-30 μg m^{-3} in Figure 6.2) where the external contribution is ~30-35%. Of the external contribution a large proportion arises from United Kingdom power stations, varying from about 20-30% in the more remote areas up to 60-70% in the regions downwind of the major power stations in the East Midlands and Yorkshire. These areas include York and Lincoln.

5.3.2 Calculations for Earlier Periods

The previous section dealt with the estimation of source contributions to SO$_2$ concentrations in 1983. In this section some estimates of contributions in earlier years are considered. This discussion should be taken in the context of historical trends in emissions and air quality described in Section 4.2. Any estimates of source attribution in these earlier years will be approximate, largely due to the lack of detailed emission inventories for any urban areas in the United Kingdom.

An earlier dispersion modelling study of SO$_2$ (and smoke) in London was carried out by WSL (Keddie & Williams 1976) for 1962 as part of a study of smog episodes in London. The emission data were calculated from fuel consumption data for the whole of London so accurate spatial disaggregation was difficult. Consequently, 10 km grid squares were used. The results averaged over the Greater London area are shown in Table 5.2. Also shown are the results of the calculations for the central 40 × 40 km zone in London for the

Winter 1975/6 and 1979/80 periods (Spanton 1983), and the WSL results for London for 1983 from Table 5.1. (The relative contributions to winter averages should be within about 5% of the contributions to annual averages since most emissions are from space heating sources).

Table 5.2 Contributions to average SO$_2$ concentrations in London 1962-1983.

	Year			
	1962	1975/6†	1979/80†	1983
Observed (μg m^{-3})*	252	128	83	35-72
Calculated (μg m^{-3})*	272	116	96	60-71
% Contributions:				
Domestic	42	8	8	
Comm/Ind+	47	64	57	67-72
Traffic	2	4	5	
Non-London Sources	9	24	30	30-33

* Averages for whole London area for 1962, for the central 40 × 40 km area for 1975/6 and 1979/80 and for a point at Westminster 6 site for 1983.

† Winter averages for 1975/76 and 1979/80.

+ Include power stations within London area, contribution from which is small, 5% of total.

The picture which emerges from Table 5.2 is one of declining absolute and relative contributions from local sources. Although the proportional contribution from sources external to London have become relatively more important, absolute concentrations have decreased by a factor of roughly five over this period. In the early 1960s (and probably before) local sources were contributing in the order of 90% of the total annual average SO$_2$ concentration. By 1983 the figure was 67-72%.

A qualitatively similar situation probably existed in most urban areas of the United Kingdom in the early 1960s. The urban contribution to the SO$_2$ concentration at Lincoln in 1962 can be estimated from the data in Table 5.1 and the fact that non-power station emissions in the United Kingdom in 1962 were 3.6 times those in 1983. The estimated figure is roughly 40-80 μg m^{-3}. This is in reasonable agreement with the measured data (Figure 5.1) which suggests an urban contribution of about 90 μg m^{-3}.

There have been few detailed dispersion modelling studies carried out on urban areas in the United Kingdom other than those already referred to. One such study was carried out by WSL in the Glasgow area in 1977-79 (Williams *et al* 1981). This showed that when averaged over the whole city area, 87% of the annual mean SO$_2$ concentrations arose from sources within Glasgow, the broad breakdown over the city area being 10% from domestic sources, 7% from traffic, and 70% from commercial/industrial sources.

5.3.3 Short Period (Episode) Calculations

Previous sections have dealt with contributions to long-period average SO_2 concentrations. However, the short-term exposure of materials to elevated concentrations is also potentially important to an understanding of buildings and materials damage. The distinction between timescales is important because, depending on the spectrum of source heights and the meteorological conditions responsible for the elevated concentrations, it is possible that relative source contributions quite different from the longer-period results could occur.

In the 1950s and 1960s the majority of episodes of high SO_2 concentration, such as the well-known 'smogs', occurred in winter periods when temperatures were low. Consequently, not only were space heating emissions high but anticyclonic synoptic conditions often prevailed. Low-level radiation inversions were present, effectively trapping pollutants in a layer some 100-300 m deep, with stable stratification and low wind speeds. Concentrations were thus able to rise and fogs/smogs formed, often stabilising the layer by radiation from the upper surface. While the absolute contributions from the various source categories under such conditions would be considerably greater than the annual averages, the relative contributions would be similar. In both cases the emissions would be into a relatively homogeneous layer. The only significant difference would be that under the episodic conditions the tall stacks (eg power stations) would probably emit above the inversion so that they would contribute a very small amount to ground-level concentrations. A calculation of the relative contributions to SO_2 concentrations for the annual periods 1962 and 1974/75 and for episodes (lasting several days) in December 1962

and December 1975 has been carried out (Keddie & Williams 1976) and the results are summarised in Table 5.3. It is clear that in these episodes there was no major difference in relative contributions other than a decrease in contribution from the tall stacks with a corresponding increase in relative contributions from the low and medium level sources.

The statistical distribution of concentrations from local sources in a city is quite different from that of concentrations from a distant elevated source. The difference arises because local sources will make some contribution whatever the stability or wind direction; the distant source can only contribute if the wind is in the right direction and if there is sufficient turbulence to disperse the plume to the ground. In practice it is found that the contribution from local sources obeys a lognormal distribution while that from a point source obeys an exponential distribution (Larsen 1969; Barry 1975). This effect is illustrated in Appendix B where it is shown that the urban contribution to annual SO_2 concentrations in York is probably a factor of three larger than that from the Drax A power station, but the ratio of peak hourly concentrations can be up to a factor three in the other direction. It should be stressed that the calculations deal with extreme value statistics and are thus approximate.

It should be noted that the above calculations are simply an extrapolation from the normal range of meteorological conditions. Truly extra-ordinary conditions which could give rise to sustained periods of high concentrations, either from urban sources (ie smogs) or from power stations (eg persistent anticyclone in South Yorkshire, sea breezes in the Thames Estuary), would not necessarily be frequent enough to be predicted by these statistical models.

In general, it may be seen that if receptor damage is caused by high concentrations over averaging times of less than 24 h, then the nearby point sources are likely to be more significant than would appear from their small contribution to the long-term average.

Table 5.3 Relative contributions to SO_2 concentrations in Greater London during annual and episode periods in 1962 and 1974/5.

| | Year | | | |
| | 1962 | | 1974/5 | |
	Annual	Episode	Annual	Episode
Observed ($\mu g\ m^{-3}$)	252	2277	95	573
Calculated ($\mu g\ m^{-3}$)	272	2480	116	719
% Contributions:				
Domestic	42	46	12	27
Comm/Ind	41	42	53	57
Traffic	2	2	3	4
Urban Power Stations	6	0	14	0
Non-London Sources	9	10	18	14

5.4 OTHER POLLUTANTS

Although SO_2 is considered to be the most important pollutant involved in damage to stonework and many other materials, other species may also contribute. A brief discussion of the emissions and air concentrations of these species has already been given in Sections 4.3-4.7. There have been few modelling studies carried out in the United Kingdom for species other than sulphur, either in urban areas or over longer distances.

5.4.1 Smoke

In the years prior to the Clean Air Act 1956, and for several years after as smoke control programmes developed, coal was the major source of combustion particulates in urban areas. The earlier WSL modelling work referred to above (Keddie & Williams 1976) suggested that in London in 1962 75% of combustion particulate concentrations arose from domestic solid fuel with a further 3-4% from industrial coal sources. Motor vehicles were estimated to contribute 11% of the concentrations. At that time total concentrations were very much higher than they are at present (Figure 4.5).

More recent figures from a study in Glasgow (Williams *et al* 1981) show that the average contributions to measured annual average black smoke concentrations in 1977/7 were 40% from motor vehicles (with ~33% from diesel vehicles) and 40% from domestic sources (chiefly bituminous coal), the remainder arising from commercial/industrial sources (5%) and sources outside the city (5%). Given that the consumption of domestic coal relative to vehicle fuels was probably higher in the Strathclyde area than in London, these results are consistent with the estimate of a vehicular contribution in London of up to 77% long term (annual) average in the mid 1970s (Ball & Hume 1983). As coal combustion decreases, motor vehicle emissions of black smoke become proportionally more important. Accurate estimates are difficult, not only because of the lack of detailed emission inventories, but also because of the uncertainty over particulate emission factors, especially from motor vehicles. Nevertheless the contribution to ambient concentrations from traffic sources in London is now likely to be of the order of 80% on average. In busy street canyons, the contribution may be even higher (Perrin 1984) and in such situations the vehicular contribution to SO_2 levels may also be significant (Bennett *et al* 1986). Finally, it should be noted that these figures take no account of possible different reactivities of particles from different fuel sources.

5.4.2 Oxides of Nitrogen (NO$_x$)

Dispersion modelling of NO and NO_2 concentrations is complicated by the reactive nature of the species involved. However, some work has taken place in the Netherlands (Van Egmond & Kesseboom 1985) using simplified

chemistry in a meso-scale model. On urban scales NO_x can be treated as a conserved species and Gaussian plume modelling studies of NO_x can be carried out using an emission inventory. Such a study was undertaken by Perrin (1984) and Simpson (1986) using the GLC inventory.

In terms of long-period averages, contributions to total NO_x concentrations in Central London for 1979/80 from motor vehicles, including sources outside London, were 60-70%. Power stations outside London were estimated to contribute some 3% of the total, with the remainder being accounted for by contributions from the commercial/industrial sector with London (12%), and other stationary sources outside London (15-25%). Model calculations for a period of relatively elevated concentrations (29-30 September 1982), suggested that proportional contributions to SO_2 and NO_x from motor vehicles are markedly different, amounting to at most 14% for SO_2 but up to 85% for NO_x.

5.4.3 Ozone

The modelling of ozone concentrations and those of other species produced photochemically in the atmosphere is complex. As has already been noted in Section 4.6, the Photochemical Oxidant Review Group report (PORG 1987) addresses the modelling work carried out to date in the United Kingdom on photochemical oxidants and the reader is referred to that report for a discussion of these issues.

5.5 CONCLUSIONS

(i) The atmospheric dispersion models discussed in this chapter are capable of providing reasonably accurate estimates of the contributions to airborne pollutant concentrations from the major sources categories. However, the calculations presented in this report used very approximate estimates of emissions in the towns studied. In order to provide improved information on the urban contribution to pollutant concentrations around historic buildings and other areas, detailed emission inventories are needed. Such information will be essential to the construction and evaluation of pollution control policies.

(ii) In an investigation of a sample of 26 towns in the United Kingdom, the external contribution to total annual

(1983) mean SO_2 concentrations for the town was estimated to be at least 40% and in most cases 50% or greater. The exceptions are those towns with a population of several hundred thousand which lie outside the higher SO_2 regions of the United Kingdom (eg those broadly outside the 20-30 μg m^{-3} annual mean), where the external contribution is 30-35%. A larger proportion of the external contribution of SO_2 now arises from United Kingdom power stations. The amount varies from 20-30% of the external contribution in remote areas, up to 60-70% in regions downwind of the major East Midlands and Yorkshire power stations.

(iii) For episodic conditions in urban areas, calculations demonstrate that low and medium level sources tend to be more important than high level sources, but under suitable conditions high level sources may dominate for short periods.

6 Mechanisms of deposition and decay

SUMMARY

The interactions between building materials and pollutants are very complex and many variables are involved. Deposition of pollutants (as particulates, aerosols, or gases) on to building surfaces depends not only on their atmospheric concentrations and the turbulence profile at the boundary layer, but on the strength and direction of the wind and the rain intensity, both in the open and immediately around the deposition site. Once pollutants are present on the material surface, interactions will vary with the position of the material (eg its degree of exposure to wind and rain). The time-of-wetness of the surface is particularly important. Finally, the natural structure and reactivity of different materials and the degree to which they are protected, will determine the degree of damage caused.

The complex character of deposition and damage is further compounded by the large number of uncertainties, particularly in reaction mechanisms and the role of different pollutants. Natural processes play a significant role in the degradation of building materials, particularly frost, wind and rain. Atmospheric carbon dioxide is thought to lead to dissolution of calcareous stone but its importance relative to other pollutants is not quantified. Some degradation of stone is probably caused by recrystallisation of salts deposited in the past (the 'memory effect'), but its effect relative to degradation initiated by present pollutants is not known.

6.1 INTRODUCTION

Mechanisms of deposition and the environmental factors that influence them have important implications for the types of material, the specific areas of buildings affected, and the forms of damage. In particular the relative importance of wet and dry deposition and the degrees of exposure will decide whether there is a net gain or loss of material and how evenly the damage occurs. These factors, in turn, relate to the protection and preservation of materials (see Chapter 9).

The interaction between pollutants and the building material on which they are deposited gives information on the relative importance of each pollutant and indicates those that are incorporated in decay products, those that form a surface crust and those that have a catalytic effect (eg combustion particles). Knowledge of all of these is required for any study of the benefits of reduced pollutant emissions.

6.2 MECHANISMS OF POLLUTANT DEPOSITION

6.2.1 Introduction

The two mechanisms for transferring pollutants to material surfaces are defined as dry and wet deposition. These terms represent simplifications when considering the transfer of material to anything other than a horizontal plane surface, particularly in urban areas where complex air flows around buildings, the complications of microclimates and driving rain can play important roles.

There is more information available on the dry and wet deposition of sulphur species (chiefly SO_2 and SO_4^{2-}) than other species and the United Kingdom Review Group on Acid Rain (RGAR) summarised the information available up to 1981 in its first report (RGAR 1983). This work has been updated in the Group's second report

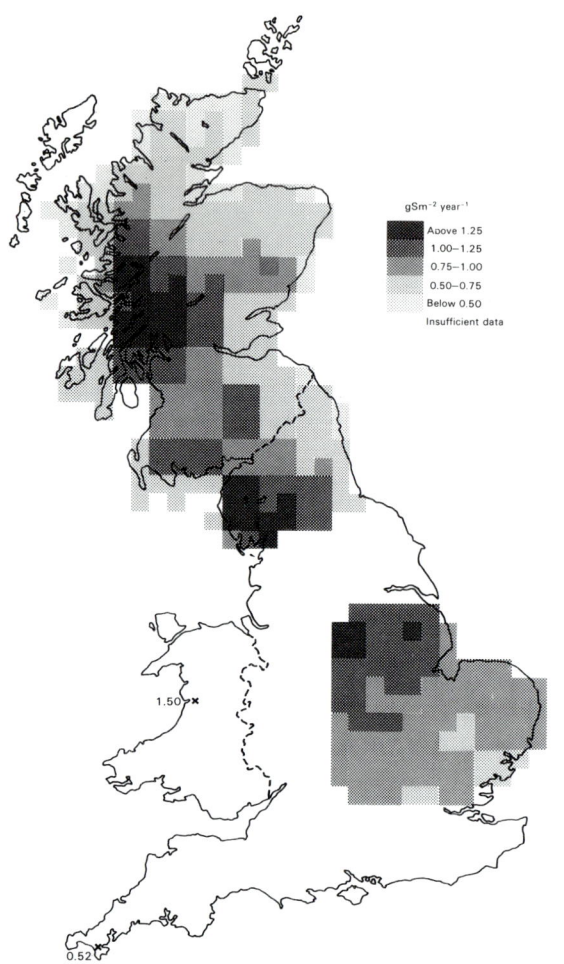

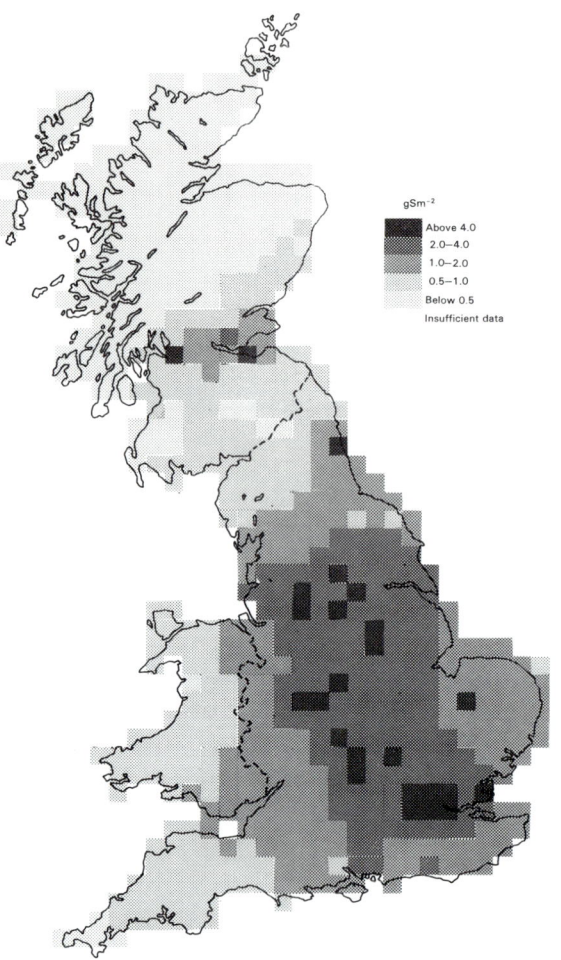

Figure 6.1 Wet deposited non-marine sulphate, 1981-1985 (based on 20 × 20 km grid squares) (RGAR 1987).

Figure 6.2 Dry deposited sulphur, 1983 (based on 20 × 20 km grid squares) (RGAR 1987).

published in 1987. Figures 6.1 and 6.2 are taken from the second report and illustrate the dry and wet deposition of sulphur in the United Kingdom. The dry deposition pattern was calculated by applying a dispersion model to airborne concentrations which were then multiplied by an average deposition velocity, based on land use data, for each 20 × 20 km grid square. The wet deposition map was derived from measurements of sulphate in rainfall, but these were not available for many areas in England and Wales.

6.2.2 Dry Deposition

The processes by which pollutants, gases or particles, are transferred from the atmosphere to solid (and liquid) surfaces are complex. The transfer eventually takes place in the boundary layer and is affected by turbulent transfer. The division may conveniently be made into atmospheric and surface processes. The atmospheric processes include an aerodynamic

resistance (r_a) which is determined by the wind velocity and aerodynamic roughness of the surfaces, and a 'bluff-body' resistance (r_b) which is a correction to take account of differences in the transfer of momentum and gases to rough surfaces (Chamberlain *et al* 1984).

The surface processes are complex and will depend on a number of variables including the nature and state of the surface. The total surface resistance to deposition can be denoted by r_s and, in analogy with electrical resistance, the total resistance to deposition is r_t and is given by:

$$r_t = r_a + r_b + r_s$$

The deposition velocity is then defined as $v_d = 1/r_t$.

In the RGAR (1983) report, a value of 17 mm s^{-1} was suggested as a typical urban deposition velocity for SO$_2$. Using an interpolation formula for the bluff-body correction (Chamberlain *et al*

1984) it appears that not only may the deposition velocity be smaller than the RGAR value but that it is also a function of wind speed. Thus for geostrophic wind speeds of 4, 8 and 12 m s^{-1} the deposition velocity over an urban area of aerodynamic roughness length 1 m would be 2.4, 4.0 and 5.4 mm s^{-1} respectively.

The value used by the RGAR is the average for urban areas, including all building surfaces and vegetation (which represents a major component of the surface in the residential areas of cities), whereas the values obtained from the bluff-body formula in the previous paragraph may be more appropriate for 'buildings' than for the entire urban area. However, both of these approaches provide only approximate values and the difference may be seen as a reflection of the large uncertainty which exists. More field and laboratory studies are required to resolve these problems.

The significance of the above range of values for the gas-phase resistance depends upon the value of the surface resistance and this is known to depend upon the physical state of the surface, in particular on the humidity and the pH.

Liss (1971) found that for SO_2 deposition to a water surface (pH 4.0), the resistance on the liquid side of the interface was negligible compared to the gas phase resistance. This would presumably also be the case for a film of rainwater on building stone. Spedding (1969) measured the uptake of SO_2 to a nominally dry oolitic limestone surface and found that at a relative humidity (RH) of 80% the surface resistance was 0.0095 s mm^{-1} (no separate measurement was made of the aerodynamic resistance of the chamber) whilst at an RH of 12% this increased to 0.7 s mm^{-1}. Payrissat and Beilke (1975) studied the uptake of SO_2 by a range of calcareous soils at various pHs and found that, for a soil of pH 7.6, the surface resistance was 0.039 s mm^{-1} at 44% RH and 0.024 s mm^{-1} at 88% RH. Similarly, in a study of a chalky field, Garland (1977) found no appreciable surface resistance (ie < 0.05 s mm^{-1}) even when the surface appeared to be dry. These measurements would thus appear to be in agreement at high humidities but the resistance of a dry surface seems to be more problematic. It is possible that soil and stone have different properties when dry, with the more fragmented structure of dry soil allowing SO_2 to penetrate further to find suitable receptor sites. Overall the results suggest that, in very dry conditions the deposition velocity to limestone is dominated by the surface resistance and the velocity may fall to 1 mm s^{-1}. In humid conditions, the surface resistance will be small (0.05 s mm^{-1}) but may be significant in high winds or exposed locations, where deposition

velocities may exceed 5 mm s^{-1}. For rain-washed surfaces, only the aerodynamic resistance should be significant.

It thus appears that aerodynamic resistance is the principal factor governing dry deposition of SO_2 in a city, the gas-phase resistance being typically in the range 0.2 - 0.5 s mm^{-1}. A precise calculation of gas-phase resistance depends on an estimate of the bluff-body correction and its value in a city relies on extrapolation from measurements with roughness Reynolds numbers (see glossary) an order or magnitude smaller. If aerodynamic resistance is the principal limiting factor, it follows that surfaces exposed to the wind will suffer higher deposition.

Although it is well established that a humid limestone surface (RH > 90%) will have a low surface resistance (< 0.005 s mm^{-1}), the behaviour of the surface at lower humidities has not been so well studied. This could be significant for estimating the total deposition in the summer months.

The dry deposition of NO_x is somewhat less well understood than that of SO_2, particularly in urban areas. Over the land surface of North-West Europe as a whole, the deposition velocity for NO is small; that of NO_2 while greater than that for NO is much smaller than that of HNO_3. Current estimates suggest values of deposition velocities of ~4 mm s^{-1} for NO_2 and ~25 mm s^{-1} for HNO_3. It should be stressed that these do not necessarily apply to urban areas and building surfaces.

In summary, there are significant uncertainties in the understanding and specification of the turbulent transfer of pollutants to urban surfaces which severely limit the accuracy with which calculations of deposition to building and material surfaces can be made. Moreover there are additional uncertainties in the nature of the variation of deposition velocities with wind speed and wetness of the surface which can further complicate source attribution.

6.2.3 Wet Deposition

Wet deposition is due to precipitation and consequently is an intermittent event with both spatial and temporal variations. Pollutants can become incorporated by two different mechanisms, in-cloud scavenging ('rain-out') and below-cloud scavenging ('wash-out') (Garland 1978).

In-cloud scavenging is the main route for wet deposition and occurs when pollutants are included in the droplets developing within a cloud. It is a very efficient removal process, the rate at which it occurs probably varies with the amount and character of the pollution and the micro-physics of the cloud.

Below cloud scavenging involves the take-up of pollution by precipitation as it falls from the cloud. It seems to be a much less important process than in-cloud scavenging. Chan and Chung (1986) suggest it accounts for less than 15% of the SO_4^{2-} and NO_3^- in precipitation.

Occult deposition, a different form of wet-deposition, may be important in some areas. This arises from impaction of fog or cloud droplets which can contain high concentrations of pollutants.

The mechanisms for the wet removal of NO_x from the atmosphere are uncertain. It is probable that gas phase transformation of NO_x to nitric acid followed by incorporation in droplets forms the main route in daylight. In darkness, the reaction of N_2O_5, which subsequently forms nitrate aerosol and is scavenged by droplets, may constitute a significant route (Richards 1983).

After the pollutants are incorporated into the precipitation, they are subject to the influences of the wind. How they are deposited onto building materials will depend on the location of the material, the intensity of the rainfall and the wind strength and direction. The interactions of wet and dry precipitation and building materials are discussed below (Section 6.3).

6.2.4 The Relative Importance of Wet and Dry Deposition

The RGAR (1983) report suggested that in areas remote from major pollution sources and where rainfall is high, wet deposition will predominate. In Eastern England, dry deposition will be the dominant process (Figure 6.3). A more precise estimate can be obtained for South-East England from the study by Jaynes (1985) which is described in Section 2.3.1 (c). Measurement of weight loss from exposed samples and leached weight loss from sheltered ones suggested that up to 40% of the total damage was due to dry deposition. It must, of course, be remembered that dry deposition will increase sample weight, that damage will only occur when moisture is present and weight loss only when runoff or physical destruction occurs. (Figures 6.9 and 6.10).

6.2.5 Factors Affecting Deposition

It is clear that both wet and dry deposition are affected by wind and rain intensity. These are particularly variable around buildings in urban

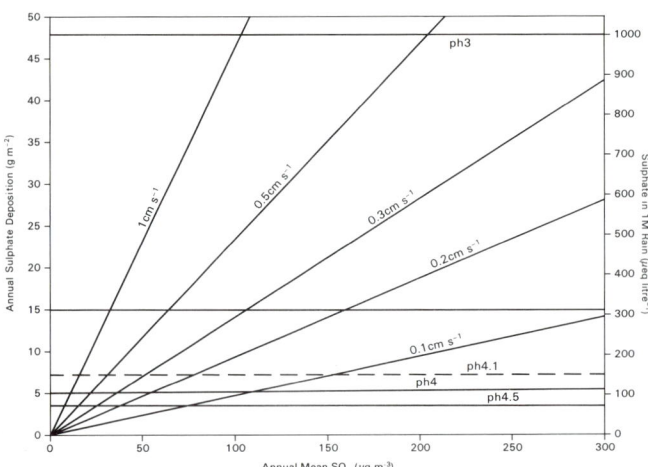

Figure 6.3 Dry deposition of SO_2, with various deposition velocities, compared with total wet deposition from a nominal 1000 mm of rain with average pH values (due to SO_4^{2-}) as shown.

areas. Precipitation will mobilise previously deposited pollutant and reaction products and it can also erode a surface by impact or by its action in flowing over the surface. Surface moisture is also important for dry deposition as it reduces the surface resistance (see above).

The action of wind may deflect rain onto horizontal and vertical surfaces. The relationship between rainfall and wind parameters is quite complex, involving semi-empirical relationships between rates of rainfall, drop-size distribution, and terminal velocities. The 'driving rain index' is a measure of the intensity of rainfall on vertical surfaces, variations of which across the United Kingdom are shown in Figure 6.4. The directional component of the driving rain index also has an important bearing on how materials are affected. This is illustrated in Figure 6.5.

It is important to appreciate that the information given in Figures 6.4 and 6.5 is purely an expression of standard meteorological measurement taken on a national scale. In practice, local conditions of topography and terrain roughness (eg surrounding buildings, fences, vegetation) significantly modify the wind direction and strength and consequently the impact zones of precipitation. Scaling factors can be applied to adjust for topography, terrain roughness and building height. The scaling factors are an expression of the way surroundings affect the wind velocity profile and wind speeds. The effect of buildings on wind flow is shown in Figure 6.6, the entire pattern of wind speed and direction being controlled by building height and spacings, the pitch of roofs, etc.

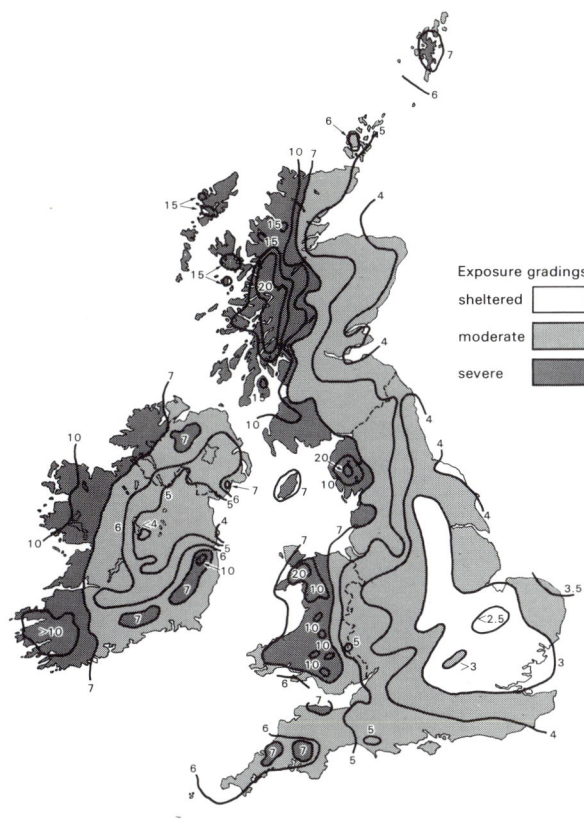

Figure 6.4 Map of annual mean driving rain index (DRI) for the British Isles (in $m^2 s^{-1} year^{-1}$).

Exposure gradings
- sheltered
- moderate
- severe

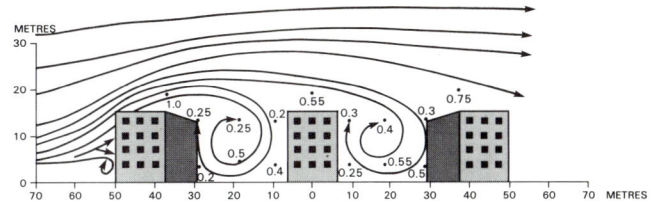

Figure 6.6 Typical wind patterns to be found between buildings at the edge of an open environment. Note: spot data is the ratio between measured windspeed and speed at the same elevation on an equivalent open site (based on wind tunnel data).

Each individual building will also modify the wind pattern in a very local sense which leads to local spatial variations in the intensity of rainfall incident on a surface (Figure 6.7). It is worth remarking that if rain drops followed wind paths exactly, no water would fall on vertical building surfaces. It is the momentum of the droplets that carries them away from wind paths and the departure is greatest where wind direction changes abruptly.

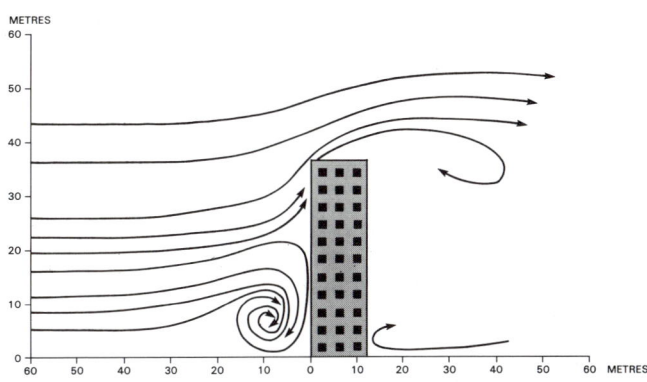

Figure 6.7 Typical wind patterns around a slab type building (from Wise *et al* 1965).

Surface features on a building can also have a pronounced effect. The sheltering effect of even small horizontal projections can be seen from wetting patterns or staining. Overhangs, such as string courses and sills, are traditional protective features for walls. There is little quantitative information available on the behaviour of water in close proximity to walls and in particular on the effect of surface texture and protective or decorative features. However, during rainfall, it has been noted that water collected and shed by such features tends to fall vertically in the air boundary, rather than being blown back onto the wall.

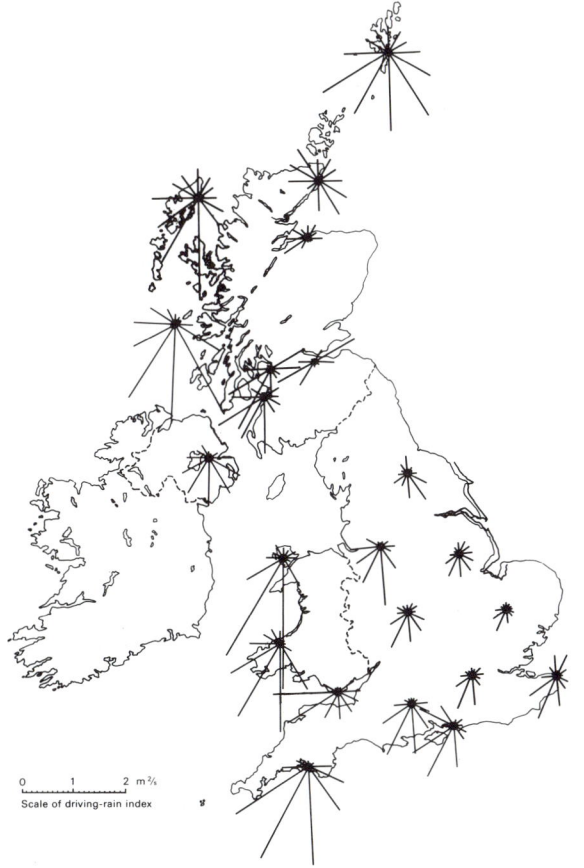

Figure 6.5 Directional components of driving rain index (DRI) measured at fully recording Meteorological Office sites.

39

In some cases, poor design features may lead to channelling or concentration of water onto certain parts of the wall. Although the surface as a whole may be quite sheltered, these specific regions will be effectively exposed and vulnerable. A common example of this is walling under windows that is not protected by an appropriate sill.

6.3 DAMAGE MECHANISMS: EVIDENCE FROM STANDING BUILDINGS AND SITE EXPOSURE TRIALS

6.3.1 Stone

(a) Natural Weathering

Porous stones are affected by frost and by impregnation with soluble salts. The detailed mechanisms by which damage occurs are not completely understood but include the expansion of water on freezing or the expansion of the salts as they crystallise and hydrate. However, simple expansive reactions are not the entire cause of damage. For example, non-reactive materials, which do not undergo expansion on freezing (eg nitro-benzene) can still have an adverse effect when the stone is subjected to freezing cycles. (Honeyborne & Harris 1958).

Frost damage causes large sections of the surface of susceptible stones to split away after cracking (Figure 6.8). The damage particularly occurs when the moisture content of the masonry is high for long periods. Salt weathering can occur naturally (Sperling & Cooke 1985) or can be caused by acid deposition. The mechanisms by which it occurs are discussed below. Both frost and salt damage are related to the pore size distribution of the stone, its chemical composition, and its mechanical properties. For example, it is known that limestones which have a high percentage of very small pores are less durable than those that do not.

The natural levels of gases in the atmosphere can, in theory, produce acid deposition and consequently degradation of building materials. Rainwater saturated with carbon dioxide can have a pH of 5.6 (RGAR 1983). Chapter 2 describes an experiment which attempted to determine the contribution of CO_2 to the decay of limestone and Chapter 4 outlines the trends in CO_2 concentrations during the 19th and 20th centuries.

The basic chemical reaction mechanisms are:

$$CaCO_3 + 2CO_2 + 2H_2O \rightarrow Ca(HCO_3)_2 \qquad (1)$$

In the presence of sulphur dioxide the reaction can continue:

$$Ca(HCO_3)_2 + SO_2 \rightarrow CaSO_3 + 2CO_2 + H_2O \qquad (2)$$

Oxidation and hydration can then take place, leading to the formation of gypsum ($CaSO_4.2H_2O$).

In the pH range 4-7, reaction (1) dominates the dissolution process but at lower pH, direct attack by H^+ ions becomes significant (Plummer et al 1979). Overall reaction rates will also be influenced by transport and diffusion processes of reactants to the surface (Charola 1987).

At present the evidence for the extent to which CO_2 is responsible for materials degradation is too incomplete for definite conclusions. However, with the prospect of increased CO_2 in the future (the 'greenhouse effect'), it is an area that will require further study.

(b) Damage on Rain-Sheltered Surfaces

Much of the damage observable on historic buildings is the formation of crusts on relatively rain-sheltered surfaces of limestone and calcareous sandstones. These are often blackened by the incorporation of soot. Spalling of the crusts causes a dramatic loss of the original surface features. The crusts, which are rich in hydrated calcium sulphate, form as a result of dry deposition of sulphur dioxide into the pores of moist stone (Schaffer 1932). Thicknesses of several mm can build up after 50 years of exposure in a heavily polluted atmosphere ($\sim 300\ \mu g\ m^{-3}\ SO_2$) (Figure 6.9).

At relative humidities greater than 80% (ie moisture present in pores) the most likely gypsum formation reaction will be:-

$$CaCO_3 + H_2SO_4 + H_2O \rightarrow CaSO_4.\ 2H_2O + CO_2$$

The sulphuric acid is produced by oxidation of SO_2. Catalysts for this latter reaction can be found in soot and dust (see below).

There is little evidence for nitrates being present in the crusts but the study at St Paul's Cathedral, London (Sharp et al 1982) showed an increase in nitrate concentration as rainwater flowed over limestone. Similar results have been observed by Leysen et al (1987). It has also been suggested (Rosenberg & Grotta 1980; Johansson et al 1986) that NO_2 enhances SO_2 pick-up in a synergistic manner without producing nitrates (see Section 6.4).

(c) Damage on Rain-Washed Surfaces

Visually, the rain-washed areas of limestone and sandstone buildings often appear to be in better condition than the rain-sheltered areas (Schaffer 1932; Blaeuer 1985) because the degraded material is being washed away continually to leave a fresh surface. Some rain-washed areas of buildings are particularly susceptible to dissolution because of the microclimate. It is then important to know how dissolution is affected by acid pollutants in the air and in the rain. Rain-washed surfaces will be damaged by natural processes such as pitting and solution weathering in drip zones. Dissolution of sulphates and nitrates will occur in zones where water is flowing over the surface. This may be accompanied by erosion of stone fragments that have been loosened by natural weathering (frost or salts) or by dissolution (Figure 6.10).

Under some conditions, deposition of sulphates and recrystallisation of calcium carbonate may also occur. Dissolution and recrystallisation of calcium sulphate within stone especially in old buildings and monuments has been termed the 'memory effect' because of the likelihood that some of the sulphate may have been deposited in the past. Several mechanisms can be postulated. For example, sulphate in solution can be drawn up from an adjacent sheltered area, which had been exposed to the atmosphere for a number of years, by permeation and evaporation and deposited in or under a rainwashed surface. If gypsum forms as the moisture evaporates, the surface may then suffer degradation. The extent of the degradation would be greater than that suffered by new stone in a similar pollution climate. However, at present, only qualitative evidence exists for this mechanism and the extent to which it occurs is not quantified.

Chapter 6.4 discusses the relationship between dissolution of marble and pH. The general conclusions are that dissolution is insensitive to pH changes but that it does depend on surface roughness and rain intensity (Guidobaldi 1981; Guidobaldi & Mecchi 1985). Some of the chemical and morphological changes which were observed on exposed stones are described in Chapter 2.

(d) Effects on Particulate Solids on Stonework

The physical and chemical characteristics of particles will influence their effects on stonework. For example, examination of marble from different locations and in various states of deterioration has resulted in the identification of two types of particles (Del Monte et al 1981):

(i) spherical shape, irregular rough surface, and high porosity (probably from oil-fired combustion plant);

(ii) spherical shape with smooth surface (probably from coal-fired combustion plant).

Both types of particles have median diameters of about 10 μm and contain carbon, silicon, sulphur, aluminium and calcium as major constituents. Particles thought to be from oil-fired plant are high in vanadium. Del Monte et al suggested that sufficient sulphur had been present in the particles themselves to account for the thickness of the gypsum layers present on the surface of the stone. An alternative view is that carbonaceous material (eg smoke and soot) may catalyse the oxidation of SO_2 to sulphates (Amoroso & Fassina 1983; Jacobson 1972; Johnson et al 1983; Novakov et al 1974). It has also been reported that SO_2 oxidation on carbon particles is enhanced by trace amounts of NO_2 and O_3 (Cofer et al 1980; 1981). The implication of carbonaceous particles from oil-firing in urban sulphation is considered in some detail by Del Monte et al (1984). They conclude that these particles are important agents in urban stone sulphation but acknowledge the need for validation of the catalytic action of carbonaceous particles.

Apart from carbonaceous material, other particulate solids such as those from metal smelters may be acidic in their own right and contain catalytic metal ions. Such materials may either participate in the generation of acidic sulphate on the stone surface or deposit sulphates directly on that surface. Although deposition of such particulate material on the surface may contribute to the rate of decay of stone, no quantitative data have been found in the literature. Further in-situ measurements of particulates near building surfaces are required.

(e) The Effect of Organic Growths on Stonework

Organic growths (including algae, bacteria and fungi) are found in association with degraded stone. The most obvious damage is caused by the growth of lichens (a symbiotic relationship between algae and fungi). Initially this causes only aesthetic damage. However the presence of lichens can result in conditions under which both mechanical (growth) and chemical (build-up of CO_2, acid secretions) damage can occur (Ciarallo et al 1985). Evidence has also been found for the presence of large numbers of bacteria on degraded stone (Lewis et al 1985; Eckhardt 1985). It can be shown that bacteria have the ability to degrade stone and increase dissolution but their exact role in stone degradation has yet to be determined.

6.3.2 Brickwork, Cement, and Mortar

There is little evidence that damage to brickwork is enhanced by industrial pollutants. Bricks in themselves are highly stable and the most common form of attack results from reaction between soluble sulphates and tricalcium aluminate, an important constituent of Portland cement. This produces calcium sulphoaluminate (ettringite) with an increase of volume which may cause surface crumbling of the mortar or may disrupt the brickwork (BRE 1971).

Although many types of brick contain soluble sulphate, damage is comparatively rare, since a number of conditions must be present together to cause a serious effect (Harrison 1981). One of the most important of these is some mechanism, such as wet/dry cycling, which will carry sulphate rapidly into the mortar. For this reason, brickwork foundations totally buried in soil seldom undergo sulphate attack. No evidence has been found that sulphate attack is more frequent in urban areas. Neither sulphate in rainwater nor sulphur dioxide in air appear to contribute greatly to the problem. Acidified rain may produce some surface erosion of mortar, leaving sand grains standing proud, but this affects appearance only and not structural stability.

6.3.3 Concrete

The most important problem affecting concrete is undoubtedly the corrosion of steel reinforcement (Muller-Golchert 1982). This results not only in loss of section in the steel and debonding of steel from concrete but also in cracking of the concrete because the corrosion product is more voluminous than the steel it replaces. The pH of new concrete paste is sufficiently high to protect the steel from corrosion. However, reaction with atmospheric carbon dioxide, which can penetrate deeply into the porous structure of the concrete, produces a deepening layer of neutral material containing calcium carbonate which has little protective effect. The most important corrosive agent is sodium chloride, derived from sea-spray or de-icing salts. There is little evidence that atmospheric SO_2 plays a significant role in corrosion of reinforcements, although there appears to be a widespread impression to this effect, particularly in the United States. It also seems unlikely that SO_2 contributes significantly to the loss of alkalinity in concrete (Engelfried & Toller 1985) compared with the much more abundant CO_2. Erosion of concrete by pollutants also occurs, but it is only a surface effect and does not affect the structure.

6.3.4 Glass

Modern and medieval glasses are usually different in composition. Modern glass contains high levels of sodium and silicon, while medieval glass contains a high level of potassium and lower levels of sodium and silicon. The two types of glass have very different durabilities when exposed in polluted atmospheres.

Visually, modern glass undergoes no change other than becoming soiled by particulates. However, the composition of the surface (300-500 μm) does change. Cox and Pollard (1977) noted that up to 80% of the sodium may be leached out, with a relative increase ($\sim$30%) in potassium, calcium and other metal ions. Alderborn (1971) made similar observations and showed that the changes only occurred in humid atmospheres and where SO_2 and CO_2 were present. He suggested that the reaction occurring was:

$$2Na^+ \text{ (Glass)} + SO_2 + \tfrac{1}{2}O_2 + H_2O \rightarrow 2H^+ \text{ (Glass)} + Na_2SO_4$$

The soluble Na_2SO_4 can then be washed away. The remaining surface is usually more durable and damage occurs to a limited depth only. However, if the damage is extensive the glass can become opaque.

Medieval glass is less durable than modern glass because the high levels of potassium ($\sim$20%) and calcium ($\sim$20%) used in its manufacture depressed the silicon content. This leads to a weakening of the network structure (Cox *et al* 1979) and removal of calcium or potassium from the surface results in a less durable structure than is found in modern glass.

Since sulphate is found in the corrosion layer and decay rates have accelerated since the Second World War, it has been assumed that the SO_2 in polluted air is the principal cause.

However, one school of thought considers that SO_2 may only be indirectly implicated in the decay of medieval glass (Newton 1982; 1986; Patzelt *et al* 1985). Both authors consider that water vapour in the air first attacks the glass to form hydroxides, which are converted first to carbonates by CO_2 and then to sulphates by SO_2. These sulphates form a hygroscopic crust on the surface of the glass which promotes attack by prolonging wetness (Frenzel 1985). Little is known of the effects of other pollutants, but NO_x may well have a similar action.

It is further suggested that the acceleration in the decay rate of glass is due to the fact that much medieval glass in the British Isles was stored in unsuitable damp conditions during the Second World War. German workers postulate a more direct role for sulphur dioxide (Fitz 1985) but both schools agree that attack is more rapid in the presence of sulphur pollution.

6.3.5 Metals

It was recognised in the 19th century that rates of corrosion of iron and steel were slower in inland rural districts than close to the sea-coast and higher again in urban and industrial areas. This was felt to be due to the effects of sodium chloride and sulphur dioxide.

Work carried out by BISRA from the 1930s showed clearly the significance of atmospheric humidity and of pollutants. The new understanding of the corrosion process stimulated the use of laboratory tests for assessing protective coatings including paint. However, it did not lead to a quantitative understanding of atmospheric corrosion and this still proves elusive. Also as corrosion tests tended to use over-severe conditions with grossly excessive additions of corrosive agents, they did not give very reliable predictions of performance.

During this period there was a growing realisation that atmospheric corrosion was an electrochemical process depending on the presence of water or water vapour, a corrosion stimulating agent and oxygen. It differs from other forms of corrosion in the thinness and transitory nature of the moisture film on the metal surface. Important contributions were made to the development of these ideas by U R Evans and W H J Vernon. The earlier literature up to about 1970 is well summarised by Evans (1960; with two supplements).

The layers of oxide produced on metals by oxidation in dry conditions usually act as effective barriers to diffusion of metal ions and oxygen, and hence have a considerable effect in protecting the underlying metal. However, when the oxide film is exposed to water, it may break down and permit a much more rapid electro-chemical reaction (Figure 6.11). The product of this reaction does not usually form a coherent barrier to further attack (Evans 1960; Rozenfeld 1972). Rates of corrosion are comparatively low even in wet conditions unless solutions of strong electrolytes are present to cause widespread dissolution of the protective oxide film. Rainwater contaminants, such as sulphuric acid, sea salt, atmospheric sulphur dioxide, hydrochloric acid and ammonium salts, can also contribute to the formation of corrosive solutions. Exposure to rain is not essential for corrosion, since condensed moisture can also dissolve corrosive contaminants formed by dry deposition. This is sometimes a problem in factories and warehouses and it may also affect building components. Also, since many of these species are hygroscopic or form hygroscopic corrosion products, liquid films may be formed in atmospheres well short of saturation with water vapour. Significant corrosion is usually considered to take place only above a critical value of relative humidity. This value varies with the metal and the corrosive substances present. Metal specimens placed inside screens which shield them from direct contact with rain, but allow access of air and dust, usually corrode much more slowly than specimens openly exposed at the same site. At very damp sites with considerable pollution by particulates, shielded specimens may, however, corrode faster because contaminants are not washed away by rain. Rates of corrosion are often still lower inside buildings, where heating reduces the relative humidity and largely prevents condensation. Oxygen is also essential to the corrosion process except in very strong acid solutions.

Rates of corrosion therefore depend on the time during which the metal is wetted (or during which the critical relative humidity is exceeded), on the nature and concentration of the pollutants present, and on the supply of oxygen. It has been suggested that, for a given 'time of wetness' (measured by electrochemical devices, or estimated from average relative humidity), the actual extent of corrosion may vary with the number of wet/dry cycles. This is because maximum rates of corrosion occur when the water film is at its thinnest, immediately before drying completely (Figure 6.12).

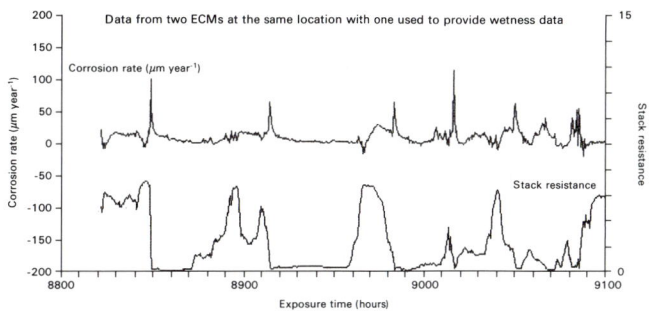

Figure 6.12 Example of continuous corrosion monitor output.

Particulates can also be important in metal corrosion. The damage they cause is related to the chemical nature of the particulate solid, the amount deposited and the nature of the metal surface (layers of oxide, hydroxide, etc, may offer some protection).

The effects of ammonium sulphate, ammonium chloride and atmospheric particulate solids on iron in the absence or presence of SO_2 has been studied (Walton *et al* 1982). The results showed that particulate material could initiate localised corrosion and increase the aggressivity of SO_2. Similar studies on aluminium, magnesium and their alloys (Olimsted 1982; Al-Ismail 1981;

Mazurkiewicz *et al* 1976) have indicated that choride-containing particles promoted pitting corrosion (localised) whereas sulphate bearing particles encourage surface thickening (non-localised corrosion).

The effects on iron of pure carbon smoke particles and hydrocarbons, with and without SO_2, have also been investigated (Johnson *et al* 1983; Skerry 1980; Arshadi *et al* 1983). Hydrocarbon gases were found to reduce the corrosivity of clean damp air. Over a period of 100-200 hours 'pure smoke', from the careful combustion of a pure hydrocarbon in clean air, was found to be more aggressive on its own than in the presence of sulphur dioxide. This finding is interesting, although it is questionable how realistic a species 'pure smoke' is.

The corrosion products that form on different metals are described in Section 2.3.3.

6.4 DAMAGE MECHANISMS: EVIDENCE FROM LABORATORY EXPERIMENTS

6.4.1 Introduction

The previous section and Chapter 2 have shown that few tests of materials have been carried out at sites with comprehensive atmospheric monitoring and even if such data were available it would be extremely difficult to analyse in terms of the effect of individual factors and damage mechanisms. Predicting the behaviour of materials in a given working environment must therefore rely to a considerable extent on data from simplified laboratory conditions in which a single factor or a small group of factors can be varied systematically while others are held constant or are ignored altogether. Other roles for laboratory studies include accelerated testing and providing standardised but artificial conditions for testing. However, artificial test environments seldom give results which can be related directly to those obtained in natural environments.

Much more laboratory research has been done with metals than with stone or other non-metals. At present, laboratory testing of stone is generally restricted to examination of 'damaged' materials by electron-optical assessment techniques for characterisation of damage. These studies are now being supplemented by experiments where factors (eg humidity, SO_2, NO_x) are assessed individually, then in binary combinations (Rosenberg & Grotta 1980; Johansson *et al* 1986). Concentrations of pollutants and temperature cycles also need to be tested in a controlled manner.

Laboratory experiments are important for establishing mechanisms of degradation, for analysing the effect of specific factors and for the construction of damage functions. Damage may be unevenly distributed in time or space (eg pitting or grain boundary attack on metals) and may thus have deleterious effects out of all proportion to the loss of material of time of exposure. The treatment and examination of specimens must take account of these possibilities. Clearly, therefore, site exposure testing and laboratory investigations in controlled environments must all be involved in characterising environmental damage to materials.

6.4.2 Metals

Corrosion testing of metals in artificial laboratory environments has often resulted in misleading results, particularly in the selection of materials and protective systems. Many of the difficulties occur because of a desire to obtain an accelerated test which tends to lead to the use of excessive levels of contaminants and periods of wetness. Difficulties also arise because it is not possible, at present, to predict the effect of experimental conditions on the balance between activation and passivation. Unexpected anomalies have been produced both by using excessive SO_2 contents and by omitting drying cycles. Laboratory tests appear however to be improved in reliability by using pollutant levels not too far above ambient values. It is becoming clear that laboratory tests using various concentrations of a single contaminant do not necessarily give a good indication of the effects which can occur in the much more complex environment encountered in exposure trials. Evidence for the importance of interaction terms has been found in the case of sulphur dioxide combined with particulates (Novakov *et al* 1974) and also with NO_2 in combination with SO_2. Johansson (1985) has found that the effect of SO_2 on iron is greatly enhanced in the presence of NO_2 at low (50%) relative humidities. Similar effects have been reported for nickel and electrical contact materials (Zakipur *et al* 1986). These experiments were carried out at rather high concentrations of contaminants, and it will be necessary to check these results at more realistic levels. Franey and Graedel (1985) have also investigated the corrosive effects of mixtures of pollutants.

It is unlikely that generally valid estimates for the effects of all the contaminants of interest and for their interaction terms will emerge from a reasonably economical series of site trials. Laboratory chamber investigations may be expected to play their most important role in the area of two and three factor interactions especially to define regions within which important synergistic effects may have to be considered in practice.

Recent interesting developments in laboratory studies of atmospheric corrosion mechanisms relate to the use of continuous electrochemical monitors. These devices usually consist of stacks of metal plates separated by thin plastic insulators, set into a block of plastic and ground down on one face to expose the edges of the plates. When a film of moisture bridges across the insulating spacers a current is produced in an external circuit, this can act as time-of-wetness meter (Figure 6.12).

Rates of corrosion can be measured from the galvanic current delivered by a bimetallic cell, or in a one-metal cell by measuring the polarisation resistance. The total extent of corrosion calculated from the integrated current in the electrochemical monitor is usually only a small fraction of that on a parallel weight loss specimen, apparently because much of the current is discharged at local cathodes rather than through the measuring circuit. A considerable body of work on electrochemical monitors has been reported by Mansfeld (1982a) and work on their use in site investigations has been reviewed by Haagenrud *et al* (1982).

A different form of monitor uses the mechanical expansion of the corrosion product in an assembly of metallic plates containing a large number of interfaces. There are many situations where structural degradation is caused by the mechanical expansion (jacking) associated with crevice corrosion which these monitors measure directly (Figure 6.12). However, their simplicity and high sensitivity also makes them useful for environmental corrosivity assessment. A technique such as metallography is used to determine the relationship between jacking and metal loss.

Both these two types of monitors give a continuous record of rates of corrosion, rather than the average value over a period given by the destructive examination of specimens from ordinary exposure trials. They have shown for instance, that the rate of corrosion is not uniform over time but may pass through sharp peaks at the beginning and end of a period of wetness. The have also shown that a large proportion of the total corrosion may take place in a short period just before the moisture film dries up altogether. Continuous monitors provide a great deal of useful information concerning the time-dependence of the corrosion process and their use in monitoring site exposure programmes appears to be very desirable. Both mechanical and electrochemical monitors can be used to estimate time-of-wetness, but they have not been used for extended periods to determine annual fractional time-of-wetness in many site exposure trials.

So far, few damage functions have been defined as a result of laboratory testing and those emerging from analysis of site trials differ greatly both quantitatively as well as in functional form.

6.4.3 Stone

Laboratory studies have a potentially important role to play in helping to resolve some of the major uncertainties regarding the mechanisms of stone decay. Amongst the areas to which they are already contributing are the synergistic effect of NO_x and SO_2 on stone, the relationship between hydrogen ion loading and dissolution, and the durability of stone. This type of study can only indicate in a predictive way what may happen in the real world, not what actually happens.

Many earlier investigations have dismissed the role of nitrogen oxides in stone decay because little nitrate is found in stone surfaces or in run-off water (Livingston 1985). However, laboratory studies (Rosenberg & Grotta 1980; Johansson *et al* 1986) suggest that NO_2 can enhance SO_2 pick-up without producing nitrates. This may now be important in urban areas since in common with many other countries United Kingdom emissions of NO_x have increased (See Section 4.4). Further investigations in the laboratory and the field are required to confirm these findings.

If acid deposition is a cause of damage to stone then it is important to understand if 'safe' levels of deposition and acidity of run-off or rainfall, exist. In a study of this subject, Reddy *et al* (1985) suggested that marble dissolution is proportional to hydrogen ion loading from the rain, but re-examination of their data showed that observed loss per mm of rain was independent of pH in the range (3.8 to 5). This agrees with the laboratory studies of Guidobaldi (1981) which showed that the rate of calcium carbonate dissolution was insensitive to pH above 4.0, though it did depend on surface roughness and rain intensity (Guidobaldi 1981; Guidobaldi & Mecchi, 1985). More research is required in order to establish the significance of rain episodes in which the pH falls below 4.0 and also on the relative effects of dry and wet deposition in stone decay.

The influence of type of mineral acid, pH and delivery rate on the dissolution of limestone in the laboratory has been studied by Bawden (1985). Although there is some variation in dissolution rate with type of limestone the susceptibility to physical damage processes is particularly variable and seems to be linked to physical properties. Work at the Building Research Establishment (BRE) (Honeyborne 1982; Leary 1983) has demonstrated large differences in durability between limestones. Even within one limestone type, such as Portland, there may be a range of durability that can be described in terms of such properties as saturation coefficient and microporosity (BRE 1983).

Laboratory studies also have an important role in determining the relative resistance of stones to natural physical weathering processes such as frost damage, salt crystallisation, and hydration stresses (Sperling & Cooke 1985).

The factors that can give rise to volume changes in stone materials have been the subject of both theoretical and laboratory studies. High pressures are calculated to be associated with salt hydration and salt crystallisation, freezing and osmosis. There are also some possible expansive reactions with gaseous species.

The study of the effects of natural physical processes and their importance relative to the effect of damage caused by pollutants is of great importance in deciding whether there is an air pollution problem and, if so, how serious it is.

6.4.4 Polymers

A large body of laboratory work has been carried out on the oxidation and photo-oxidation of many types of polymer and on reactions with SO_2, O_3, NO_2 and hydrogen chloride. Laboratory work, intended to investigate mechanisms, reaction kinetics, changes in molecular weight distribution, cross-linking and in some instances changes in mechanical properties, has in the past almost always used unrealistically high concentrations of reactive gases. This makes it difficult to apply laboratory results to the performance of polymers in normal atmospheres. As with corrosion of metals, laboratory results must therefore be considered alongside information obtained from exposure trials in natural conditions and such information is very scanty for most polymer-based building products.

The chemistry of polymer degradation is beyond the scope of this review, and no attempt will be made to comment in detail on the extensive literature. Reference may be made to the books by Davies and Sims (1983) and by Grassie (1977-1984).

Little systematic work appears to have been carried out to define the contribution of industrial pollutants to these natural degradation effects. No reports have yet suggested that either structural materials, or sealing and jointing compounds deteriorate more rapidly in industrial atmospheres than in rural areas.

6.4.5 Future Role of Laboratory Studies

Laboratory testing allows systematic study of selected variables in closely controlled conditions but it must not be regarded as exactly simulating outdoor exposure tests. Controlled conditions have indicated synergistic effects between SO_2 and NO_2 for several metals and contributed to a mechanistic understanding of the degradation of different materials. Other studies have used high pollutant levels for accelerated testing. The results allow materials to be ranked by durability, but the degradation mechanisms may not be representative of those occurring in environmental exposures.

Damage functions have sometimes been reported from laboratory studies and are usually based on the inter-dependency of a few variables under specific laboratory test conditions.

There is clearly a continuing need for further laboratory studies under controlled test conditions. These should be designed to supply more information on pollutant synergisms (eg NO_2 and SO_2) at realistic levels and allow determination of the mechanisms for the greater aggressivity of mixtures and the range of materials affected, improved damage functions, and investigation of the relative phasing of changes in pollutant concentrations, pH, etc. For stonework, the relative importance of physical and chemical weathering needs laboratory investigation so that factors such as wet/dry exposure differences can be understood.

6.5 CONCLUSIONS

The interactions between building materials and pollutants are very complex and many variables are involved (Livingston & Baer 1983). Deposition of pollutants (as particulates, aerosols, or gases) onto building surfaces depends not only on their atmospheric concentrations and the turbulence profile at the boundary layer, but on the strength and direction of the wind and the rain intensity, both in the open and immediately around the deposition site (microclimate). Once pollutants are present on the material surface, interactions will vary with the position of the material (eg its degree of exposure to wind and rain). The time-of-wetness of the surface is particularly important. Finally, the nature, structure and reactivity of different materials, and the degree to which they are protected, will determine the degree of damage caused.

In urban areas of the United Kingdom the turbulent transfer of pollutants, particularly SO_2, by dry deposition appears to be the most important mechanism for transferring pollutants to the surfaces of buildings and materials. However, very few measurements of wet deposition or of deposition via fogs and mists have been carried out in urban areas. Furthermore significant gaps exist in our understanding of turbulent transfer in built-up environments. Deposition velocities in urban areas are thus at present very uncertain, severely limiting the accuracy with which calculations of deposition of pollutants to building and material surfaces can be made. Moreover there are uncertainties in the nature of the variation of deposition velocities with wind speed and wetness of the surface. These can also complicate the source attribution estimates discussed in Chapter 5.

The complex character of deposition and damage is further compounded by the large number of uncertainties, particularly in reaction mechanisms and the role of different pollutants. Natural processes play a significant role in the degradation of building materials, particularly frost, wind and rain. Atmospheric carbon dioxide is thought to lead to dissolution of calcareous stone but its importance relative to other pollutants is not quantified. However, it is clear that calcareous stones (limestone, marbles, and calcareous sandstones), iron steel and zinc, and glass (with a high potassium, low silicon content) are at greater risk than other building materials. It is also clear that water, Cl^-, and SO_2 (and associated salts) are important in accelerating damage relative to 'natural' levels. Some degradation of stone is probably caused by recrystallisation of salts deposited in the past (the 'memory effect'), but its effect relative to degradation initiated by present pollutants is not known. Other pollutants (such as NO_x and particulates) are probably also important, but at present the evidence against them is less conclusive.

Figure 2.3 Differential weathering between Portland stone and lead plugs, St Paul's Cathedral, London (used as a measure of stone weathering).

Figure 2.1 North-East roof court and facade wall, St Paul's Cathedral, London.

Figure 3.2 An example of a traditionally constructed building.

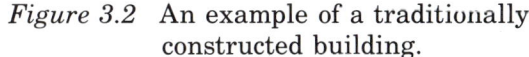

Figure 3.3 An example of a building constructed using modern techniques and materials.

Figure 6.9 An example of damage found on a rain-sheltered stone surface.

Figure 6.8 An example of stone damaged by frost.

Figure 6.10 An example of damage found on a rain-washed stone surface.

Figure 6.11 An example of metal corrosion in the atmosphere.

Figure 7.3 Lincoln Cathedral, one of the sites in the National Materials Exposure Programme.

Figure 7.4 The sample racks and monitoring equipment on the National Materials Exposure Programme site at Lincoln Cathedral.

Figure 9.1 Protection of metal. Comparison of the effect of atmospheric exposure on protected and unprotected metal surfaces.

Figure 9.2 Preservation of stone. The lower part of the sample has been treated with Brethane, and then the whole sample subjected to a crystallisation test.

7 Predicting damage

SUMMARY

Currently available damage functions do not seem to perform well in predictive use and in many cases are unsatisfactory. The main reason is that available damage functions do not take account of the complex relationships of environmental variables (eg chloride, NO_x/SO_2 synergism, rain intensity). The poor quality of data on damage occurring to building materials and the mechanisms by which the damage occurs, all of which is essential for the formulation of damage functions, has led the BERG to recommend the setting up of a National Materials Exposure Programme before publication of the main report. The main aim of this programme, which is now operational, is to determine the rates of degradation of materials in the current United Kingdom pollution climate.

7.1 INTRODUCTION

Rates of corrosion of steel measured at sites throughout the world differ by a factor exceeding 200 and a fourfold variation was found in the 1930s even within the relatively temperate atmospheric conditions of the United Kingdom. Comparable variations have also been found for zinc. It is therefore of considerable importance to develop predictive equations for calculating rates of corrosion for any given site from a set of climatic data. Such predictive equations will have an obvious application in selecting materials, in specifying protective schemes and maintenance schedules, and increasingly in assessing the cost-effectiveness of pollution abatement policies. It must, however, always be borne in mind that details of design in buildings will modify the environment of the material compared with open exposure of the simply-shaped specimens normally used in site trials. This and other complicating factors make it impracticable to seek too much precision from predictive equations.

7.2 RELATIONSHIPS BETWEEN ATMOSPHERIC POLLUTANT CONCENTRATIONS AND PHYSICAL DAMAGE TO MATERIALS

Although there is considerable qualitative understanding of the factors that determine rates of corrosion it is by no means easy to develop satisfactory quantitative expressions for predicting rates. Rates of corrosion depend upon a number of factors related to wetting of the surface (eg 'time of wetness' as variously defined, duration of rain, number of days with rain, rain intensity, average relative humidity, time at relative humidity > 80%, etc), the supply of gaseous pollutants (SO_2, NO_2, HCl, NH_3, O_3, etc), the supply of pollutants in solution in rainwater (Cl^-, SO_4^{2-}) pH, ammonium salts, organic acids, etc, and various particulate materials. Other relevant factors are temperature and the time during which water in contact with the surface is frozen. Rates vary considerably between those for free exposure to the weather and exposure inside a screen that excludes rain while allowing condensation and access to dust and gaseous contaminants. Conditions inside buildings constitute a further category. For many components of buildings (eg fasteners and reinforcements) none of these environments is entirely realistic. It therefore seems clear that the development of predictive equations must start from a wide range of measurements on the composition of the atmosphere and of rainwater, together with various humidity-related quantities. Separate equations are likely to be required for different conditions of exposure.

Rates of corrosion of many metals, including steel, are time-dependent, and the time-dependence varies between sites. Therefore it may be necessary to include a kinetic term in the predictive equation.

The usual procedure is to seek correlations between measured rates of corrosion and measured values of atmospheric parameters. To produce generally applicable equations the database should be as extensive as possible. The range of measurements at each site and the choice of sites should cover the largest possible range of atmospheric conditions. The criteria for choice of sites are clearly different from those for surveys of air chemistry or rainwater acidity. It will be necessary to use sites in both industrial and marine environments. Since there are large local variations in pollutant levels in these conditions, monitoring points need to be very close to the specimens.

The standard techniques for estimating the extent of corrosion involve destructive examination and it is impracticable to measure rates by this means over very short periods of time. For this reason, correlations are sought between rates averaged over comparatively long periods (3, 6 or 12 months) and averaged environmental data. However, this procedure does not allow detection of transient effects (episodes). The average values of many atmospheric variables tend to be strongly correlated one with another. Thus a single parameter may appear in the final equation as a proxy quantity for several others and consequently the equation may fail to predict results at different sites. It is therefore important that both atmospheric variables and rates of corrosion should be subjected to some form of continuous monitoring wherever possible. One without the other is not satisfactory.

Few trials have so far been carried out in which all the factors believed to be relevant to atmospheric corrosion have been monitored. The correlation equations therefore inevitably involve only such quantities as happen to have been measured, or those showing a statistically significant effect. In many cases the form of the final equation arises from the initial decisions on the design of the experiment. Also a number of published 'damage functions' result from later authors fitting curves to data from experiments which were not designed for this purpose.

Most of the published equations actually involve only an average SO_2 concentration and one or other of the possible humidity-related terms. Some are derived from databases that do not even include measurements of Cl^-, NO_2, or O_3. The precise humidity-related term used varies considerably from one investigation to another.

Many questions of considerable importance in discussing the effect of industrial pollutants are not resolved in investigations of this type. For example, whether the pH of rain has an important effect in its own right or merely because it happens to be correlated with SO_4^{2-} and whether limited rainfall with frequent deposition of dew is more damaging than infrequent, prolonged rain which tends to wash away pollutants. These uncertainties, inherent in a methodology based on correlations between average quantities derived from a selective database, account for a good deal of the scatter in results from different predictive equations. The situation will be improved only by more research using laboratory measurements in controlled atmospheres to:-

(i) resolve some of these qualitative uncertainties and

(ii) define the effects and interactions of pollutants which have not yet been adequately characterised from exposure trials.

A number of damage functions available for building materials have been assembled in various comprehensive compilations, some are listed below:

> Nriagu (1978);
> Umweltbundesamt (1980).
> UNECE-Airborne Sulphur Pollution (1984).
> Jorg, Schmitt and Ziegahn (1985).
> UNECE Report ENV/EB/R16 (1982).

These reviews show that fifteen or so different functions are available for steel and a similar number for zinc or galvanised steel. The smaller number of functions for stone and paint reflects the longer history of metal studies. Jorg *et al* (1985) assembled 998 references on materials damage due to air pollution and besides listing the damage function therein, provide a cross reference system for finding the relevant references for each material-pollutant combination. In the following sections, the range of functions available for major materials are considered with the aim of considering their consistency, usability and suitability for assessment purposes.

7.3 METALS

Benarie and Lipfert (1986) have published a relationship for the linear loss rate of zinc. The greater number of terms (SO_2, Cl^-, time of wetness) makes it a suitable example for extended discussion.

Loss rate (A)
= 4.8 +0.53 (SO_2 + Cl^- surface deposition, mg m^2 d^{-1})
Units of A = weight loss (gm^{-2}) per wetness year (t_w, fractional time of wetness).

Benarie and Lipfert suggest that surface deterioration is roughly proportional to air concentration, but surface properties and aerodynamic factors must also be important factors. They state that the predictive error of this equation is less than ±20% for half the cases considered in constructing the function and ±50% for the rest. If this predictive accuracy, which can be assessed by comparing their function with other observations, is correct then the function could be considered satisfactory. However, comparison of a number of functions (Table 7.1, Figure 7.1) shows that their claim is true only for their data and demonstrates the specific nature of the functions.

This equation is derived from more complex functions that also include a time-dependence term (Benarie & Lipfert 1986). These have the general form $M = a.t^b$ where the exponent b takes values between 0.5 (if the rate is controlled entirely by diffusion through a corrosion product and 1.0 (for reactions unaffected by the product). Benarie and Lipfert suggest that for atmospheric corrosion the relevant time is the 'time of wetness'. The equation then becomes:

$$M = A\, t_w{}^b$$

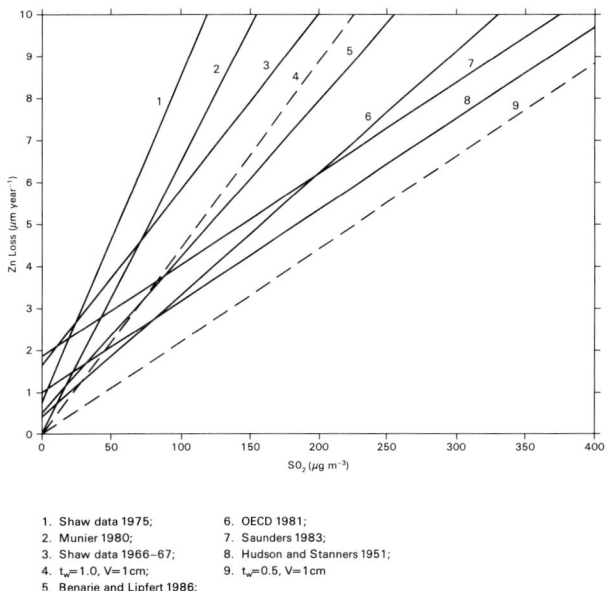

1. Shaw data 1975;
2. Munier 1980;
3. Shaw data 1966–67;
4. $t_w = 1.0$, $V = 1$ cm;
5. Benarie and Lipfert 1986;
6. OECD 1981;
7. Saunders 1983;
8. Hudson and Stanners 1951;
9. $t_w = 0.5$, $V = 1$ cm

Figure 7.1 Comparison of a range of damage functions for zinc. Note: time of wetness (where included) is taken as 0.5 and values for typical United Kingdom conditions have been used in Saunders 1983. The chloride term in Benarie and Lipfert 1986 has been set to zero.

Benarie and Lipfert point out that t_w is highly correlated with the number of days with precipitation and also with annual average humidity. It is suggested that atmospheric pollution affects rates of corrosion through an effect on factor b and that rough correlations are available between different measures of atmospheric pollution, such as hydrogen peroxide bubblers and lead dioxide candle methods for SO_2.

For steel and zinc, these relations lead to damage functions involving atmospheric SO_2 (or $SO_2 + Cl^-$) (for factor A) and average rain pH (for factor b). The data are less convincing for steel than for zinc. This approach is interesting as a means of summarising a mass of published results. However, Benarie and Lipfert point out, although the correlation equations are statistically significant, the predictive value is poor and the results lack fundamental significance. This is due to the many complicating factors and the quantities appearing in the equations do so as surrogates for other unmeasured quantities.

Edney *et al* (1986) conducted short term laboratory experiments with sub-ppm concentrations of SO_2 to determine the relative importance of wet and dry acidic deposition on the corrosion of galvanized steel. Dew was produced periodically on panels which were sprayed later with simulated rain solutions. The results show that SO_2 will deposit a fresh dry surface only until monolayer coverage is attained and

Table 7.1 Comparison of a range of damage functions reported for zinc

Study	Zinc Loss (μm exposure year^{-1})
Benarie and Lipfert (1986) (assuming 0.5 fractional time of wetness)	$0.34 + 0.037\,(SO_2) + 0.037\,(Cl^-)$
Shaw Data (1975 exposures)	$0.7 + 0.074\,(SO_2)$
Shaw Data (1976–1977 exposures)	$1.4 + 0.043\,(SO_2)$
Hudson and Stanners (1956) (1951 exposures)	$0.885 + 0.022\,(SO_2)$
Munier *et al* (1980)	$0.01 + 0.063\,(SO_2)$
OECD (1981)	$0.344 + 0.029\,(SO_2)$
Saunders (1983) (from Shaw Data for Average UK conditions)	$1.806 + 0.022\,(SO_2)$
Haynie (1980)	$2.36 + 0.022\,(SO_2)$
Atteraas and Haagenrud (1982) (assuming 0.4 fractional time of wetness)	$1.58 + 0.078\,(SO_2)$
Mikailovsky (1982) (assuming a temperature of 15°C)	$3.92 + 0.067\,(SO_2)$

Notes: SO_2 concentrations in μg m^{-3} year^{-1} (except Haynie)
Cl^- expressed as mg m^{-2}d^{-1}

deposition on a wet surface depends on the gas phase resistance of the atmosphere. For dry deposition, one atom of zinc reacts for each SO_2 molecule absorbed. Thus damage functions can be deduced from the deposition velocity of SO_2 and the time of wetness. Lines drawn on Figure 7.1 for a 1 cm s^{-1} deposition velocity for fractional times of wetness of 1.0 and 0.5 embrace many of the reported damage functions but represent unrealistic conditions. There may also be an effect caused by NO_2 which is discussed later.

In the same study, the effects of subsequent 'rain' events depended upon the earlier SO_2 concentrations. A larger 'rain' effect was attributable to lower levels of SO_2 (over the range studied 8-126 μg m^{-3}). An explanation of an effect of prior exposure conditions was suggested to explain the different corrosion products formed at low and high SO_2 levels. The effect is clearly not yet fully understood but these laboratory data indicate that when two processes operate, the dependence of zinc loss on SO_2 is not so clear cut.

Another laboratory study, Johansson (1985), has investigated the synergisms between NO_2 and SO_2 during the laboratory corrosion of steel, copper, aluminium and zinc. For zinc a synergism was detected, with 0.2 ppm NO_2 doubling the rate of corrosion obtained with 1.3 ppm SO_2 at 90% RH.

Two laboratory studies have therefore indicated that SO_2 level is not the only factor important for zinc corrosion and that other factors may have a role in explaining why zinc corrosion rates have not fallen over the years along with SO_2 levels.

Some of the spread of functions in Figure 7.1 is no doubt due to uncontrollable variations (eg in wetness parameters) which would be normalised in a more elaborate analysis. Conditions pertaining during the first two days of exposure may also cause differences (Guttman & Sereda 1968). Part of the study showed that at sites where the SO_2 content varies little over time, rates were mainly determined by 'time of wetness' during the first months of exposure and that total corrosion in the first year varied considerably according to the starting date. Rates are also strongly affected by the SO_2 content at the beginning of exposure, so that a quite different pattern would arise at sites where SO_2 content showed a seasonal dependence. Other factors such as Cl^- deposition may also affect rates, but this term appears in few damage functions. If Cl^- deposition is from sea-salts, which in common with many other particles are deliquescent, then surface wetness will be enhanced.

An extensive survey was carried out in the United Kingdom by Shaw (1978) between 1969 and 1973. Data for the corrosion of zinc at 2195 sites was collected and led to the publication of a 'corrosion map' which assigned a typical rate of corrosion to each 10 km square of the National Reference Grid. Atmospheric monitoring was carried out at 165 of the sites. Saunders (1983) produced a damage function based on the Shaw data using five variables (windspeed, temperature, rainfall, relative humidity, SO_2 concentration) in various combinations and interpolating the parameters to sites that were not fully monitored. The final equation was:

$$\text{Rate} = 1.93 - 11.96 \; I^2 + 0.0031 \; SO_2 + 0.178 \; I SO_2$$

Where I = Amount of Rainfall (average total monthly)/Duration of rainfall (average yearly mean)

Rate = μm year^{-1}
SO_2 = concentration SO_2 (μg m^{-3}, average over the exposure period)

The function includes terms that reflect the role of water in corrosion and in removing deposited pollutants.

When all these studies are considered together, there is clearly a proportionality between zinc corrosion rates and SO_2 levels but also there is no indication that the zinc losses have actually fallen as SO_2 concentrations have been reduced.

There is some doubt whether it is realistic to assign a representative rate of corrosion to a 10 km square because of large local variations in pollutant levels. Hache (1962) found a fourfold variation between 11 sites all within about 1.5 km at Biarritz. The Shaw Corrosion map (1978) bears a marked similarity to the maps of emissions and of dry deposition of sulphur prepared by the Warren Spring Laboratory (RGAR 1983), with some exceptionally high rates of corrosion at certain coastal sites. It may, however, in the future, be possible to assign average rates for 10 km squares on the basis of average pollutant concentrations derived from atmospheric modelling if satisfactory damage functions become available.

A few damage functions have been derived from databases which contain an extensive range of atmospheric measurements, notably those of Mansfeld (1980) and Haynie et al (1976). These studies both found that relatively few of the wide range of parameters monitored produced a significant effect on rates of corrosion and no significant effect was found for NO_x. In view of recent studies on synergistic effects (Johansson 1985; Zakipur et al 1986), the applicability of this conclusion to British conditions needs to be investigated.

It is clear that the available damage functions for zinc are not capable of reliably predicting actual rates of corrosion. The main reason for this is that the damage functions do not take account of the complex relationships of environmental variables. In particular chloride levels and the synergistic relationship of NO_x and SO_2 need to be included.

7.4 NON-METALS

There are few reported damage functions for stone and hardly any for other non-metallic structural materials.

Three damage functions for stone are listed by the UNECE - Airborne Sulphur Pollution Report (1984). Of these, two are from laboratory studies. One is quoted in terms of stone physical properties and another, due to Haynie et al (1976) based on a study of marble is:

$$\text{loss } (\mu\text{m year}^{-1}) = -3.31 + 0.078\% \text{ RH} + 2.95 \times 10^{-3} (SO_2) \\ (\mu\text{g m}^{-3}) \\ + 3.33 \times 10^{-3} (O_3) \\ (\mu\text{g m}^{-3})$$

The function is dominated by 'non-pollutant' terms (particularly relative humidity). Loss of material will be $\approx 0.5\text{-}4.5 \mu\text{m year}^{-1}$ in an SO_2/O_3 free atmosphere and less than $1.0 \mu\text{m}$ higher in an extremely polluted area, well below observed values (Chapter 2). These rates of degradation and the absence of terms for NO_x or time dependence, combined with the complexities of damage processes for non-metallic materials (outlined in Chapter 6) make it unlikely that this damage function would be applicable to urban environmental exposure.

A few damage functions have been published for paint. These tend to confirm that the effect of SO_2 is not very great compared with weathering in a clean atmosphere. A linear function derived for the OECD investigation from data published by Fink (Fink et al 1971) suggests that the life of paint on plain steel would be reduced by about 15% if the SO_2 content of the air increased by 100 $\mu\text{g m}^{-3}$. For galvanised steel the corresponding decrease would be about 20%. The annual average SO_2 content in British cities is now well below this concentration. The type of paint is not specified, but it appears that the data represented a cross-section of the types of paint in use in the USA. Other damage functions suggest similar small effects. The function derived by Haynie et al (1976) from a laboratory study suggests that for 'oil-based house paint' on aluminium an increase in SO_2 by 100 $\mu\text{g m}^{-3}$ would reduce the life by less than 10%. For steel, coil-coated with vinyl paint, at 100% RH an increase in SO_2 by 100 $\mu\text{g m}^{-3}$ would increase the loss by only about 6.5%. The evidence from the few available studies is not very consistent, some studies (Haynie et al 1976; Mansfeld 1980) finding a small positive effect for SO_2 and no effect for nitrogen oxides or ozone, while others found positive effects for ozone or nitrate or sulphate in rainwater. These tests covered a variety of paint materials, and it is likely that different paints react in different ways.

7.5 THE NATIONAL MATERIALS EXPOSURE PROGRAMME (NMEP)

The preceding sections show that few damage functions are available for stone, and none of these appears to be satisfactory. Many damage functions have been produced for metals, but none are of general predictive use. It is also clear from Chapters 2 and 6 that the damage mechanisms and the physical manifestations caused by the present pollution climate are, at best, only partially understood. In consequence BERG at an early stage in its deliberations suggested the establishment of a national monitoring network to investigate the decay of new building materials in the United Kingdom. The setting up of the NMEP was considered to be of the greatest importance by the government and has, therefore, already been started (April 1987). The network consists of 29 sites (Figure 7.2) at which samples of different building materials will be exposed for at least four years. Information on meteorological conditions and atmospheric pollution will be collected from all sites during this period. Four of the national sites are also included in an international network that is being operated concurrently for United Nations Economic Commission for Europe (UNECE).

The main strategic factors involved in the decision to initiate the National Materials Exposure Programme (NMEP) were:

(i) Recent expressed public concern over potential effects of acid rain on building materials within the United Kingdom (House of Commons 1984) and the apparent lack of evidence about such effects.

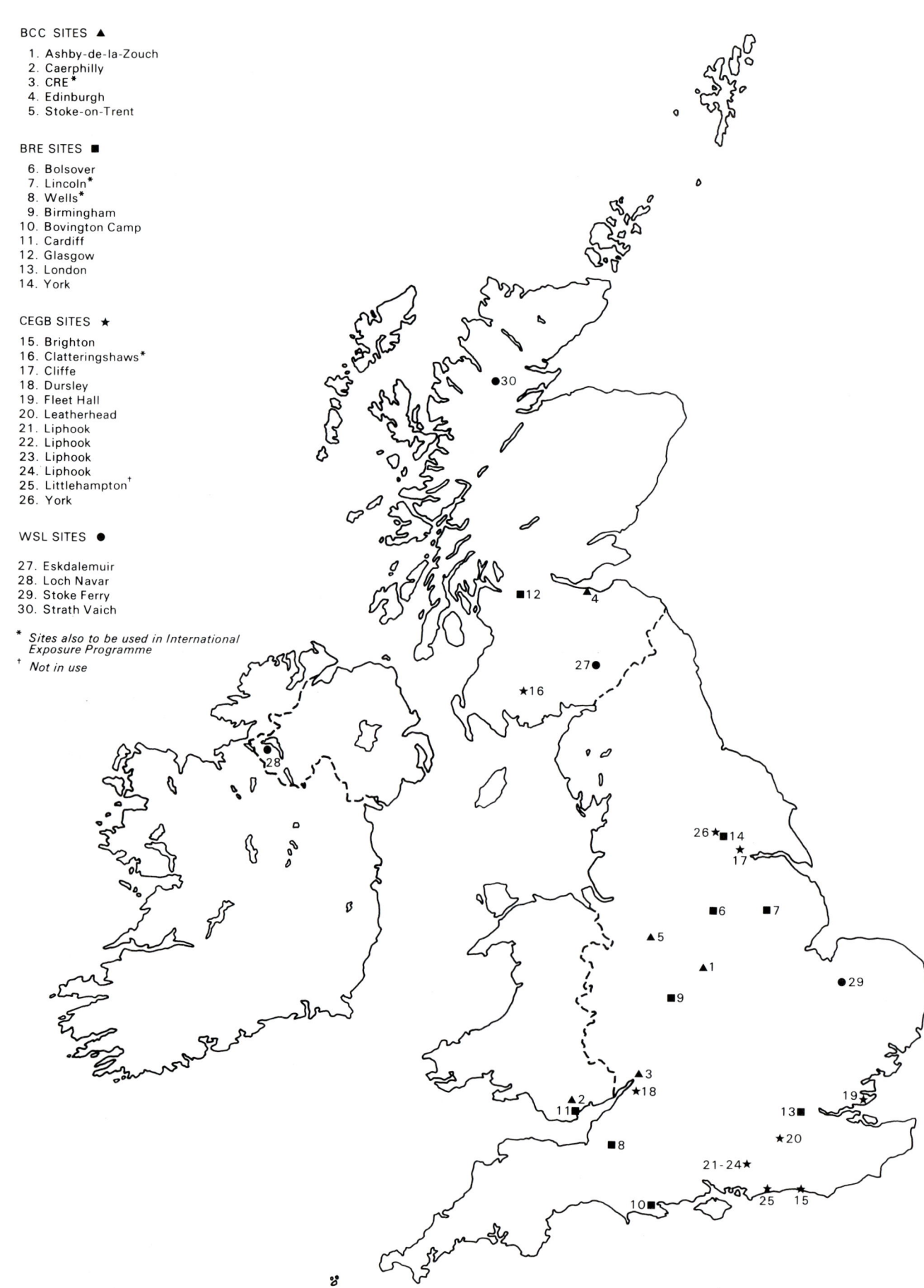

BCC SITES ▲

1. Ashby-de-la-Zouch
2. Caerphilly
3. CRE*
4. Edinburgh
5. Stoke-on-Trent

BRE SITES ■

6. Bolsover
7. Lincoln*
8. Wells*
9. Birmingham
10. Bovington Camp
11. Cardiff
12. Glasgow
13. London
14. York

CEGB SITES ★

15. Brighton
16. Clatteringshaws*
17. Cliffe
18. Dursley
19. Fleet Hall
20. Leatherhead
21. Liphook
22. Liphook
23. Liphook
24. Liphook
25. Littlehampton†
26. York

WSL SITES ●

27. Eskdalemuir
28. Loch Navar
29. Stoke Ferry
30. Strath Vaich

* Sites also to be used in International
  Exposure Programme

† Not in use

Figure 7.2 Location of sites in the National Materials Exposure Programme.

(ii) The limitations of previous work in that some was solely laboratory-based, other studies concentrated on one or two areas only and other work considered one type of material only.

(iii) The signing by the United Kingdom of the UNECE Convention on Long Range Transboundary Air Pollution with subsequent commitment to the follow-up international cooperative monitoring programmes set set up by UNECE, including materials.

(iv) The need to acquire quantitative data with which to assess the costs of pollutant damage to buildings and to apply cost-benefit and decision analyses relevant to decisions on emission abatement strategies.

(v) The need to relate laboratory-based damage functions for the effects of a variety of pollutants on building materials to data collected under actual weathering conditions on site.

(vi) Co-ordination of the research activities of individual organisations in the context of material damage so that results are cumulative and mutually comparable.

The national network is designed to provide general information upon the current rates of decay of commonly used building materials within a representative range of environmental conditions in the United Kingdom. It is hoped such information can then be used to attempt to explain observed differences in terms of climatic and atmospheric pollution conditions. The network may point to areas or particular materials which require more detailed study in the future.

The basic questions that the programme should answer are:

(a) What are contemporary rates of damage to various building materials in the United Kingdom?

(b) How do rates of decay of different building materials vary across the United Kingdom?

(c) Can the spatial variations in rates of decay of the different building materials be correlated with differences in any of the measured environmental parameters (eg rainfall, SO_2 level, etc)?

The National Site Network

The NMEP network consists of 29 sites in the United Kingdom (Figure 7.2; Figure 7.3; Figure 7.4; Table 7.2). The sites have been chosen to cover a wide range of environmental conditions, climate and topography, but gaps in the distribution exist because of the absence of suitable sites. The project is planned to last at least four years and samples of the exposed material will be removed after 1, 2 and 4 years. The materials at each site are given in Appendix C.

The International Site Network

Four sites from the National Programme (Lincoln, Wells, Stoke Orchard and Clatteringshaws) form the United Kingdom contribution to the International Cooperative Materials Exposure Programme set up by UNECE (started September 1987). The monitoring undertaken for the NMEP is sufficient for the international programme, but extra materials are included. Twelve other countries are currently taking part in the UNECE programme. The sites are listed in Table 7.3.

7.6 CONCLUSIONS

Currently available damage functions do not seem to perform well in predictive use and in many cases are unsatisfactory. The main reason for this is that available damage functions do not take account of the complex relationships of environmental variables (eg chloride, NO_x/SO_2 synergism, rain intensity). One approach is to apply a damage function relating overall rates of corrosion to pollutant levels, and possibly a wetness parameter, to area averages for the relevant pollutants obtained from deposition models. The results can then be applied to estimates of the stock at risk for various materials obtained from survey evidence or modelling (see Chapter 8). This approach was adopted by the OECD in their 1981 survey. Damage functions useful for this purpose are very simple and involve only pollutants for which satisfactory deposition models are available. The OECD calculation used a rather coarse grid for deposition and since both pollutants and materials at risk will tend to be concentrated in large towns, the averaging procedure probably underestimates the total damage. The poor quality of data on damage occurring to building materials and the mechanisms by which the damage occurs, all of which is essential for the formulation of damage functions, has led to the setting up of a National Materials Exposure Programme. The main aim of this programme is to determine the rates of degradation of materials in the current United Kingdom pollution climate.

Table 7.2 Sites in the National Materials Exposure Programme.

SITE	GRID REF.		SITE	GRID REF.	
Ashby de la Zouche	SK	391175	Brighton	TQ	253048
Caerphilly	ST	163864	Clatteringshaws	NX	553778
CRE (Stoke Orchard)	SO	980280	Cliffe	SE	662337
Edinburgh	NT	182728	Dursley	ST	755966
Stoke-on-Trent	SJ	883533	Fleet Hall	TQ	896893
			Leatherhead	TQ	155577
Bolsover	SK	471707	Liphook	SU	855299
Lincoln	SK	980718	Littlehampton	TQ	040030
Wells	ST	551459	York	SE	603524
Birmingham	SP	054861			
Bovington Camp	SY	836886	Eskdalemuir	NY	234028
Cardiff	ST	171793	Lough Navar	H	065545
Glasgow	NS	541638	Stoke Ferry	TL	700988
London	TQ	320791	Strath Vaich	NH	347750
York	SE	607514			

Table 7.3 Sites in the International Programme (UNECE).

COUNTRY	SITE	COUNTRY	SITE
Czechoslovakia	Letnany	Norway	Oslo
	Kasperske Hory		Borregard
	Kopisty		Birkenes
Finland	Espoo	Sweden	Stockholm (South)
	Ahtari		Stockholm (Centre)
	Helsinki-Vallila		Aspvreten
Federal Republic	Waldhof-Langenbrugge	United Kingdom	Lincoln Cathedral
of Germany	Aschaffenburg		Wells Cathedral
	Langenfeld-Reusrath		Clatteringshaws
	Bottrop		Stoke Orchard
	Essen-Leithe	Spain	Madrid
	Garmisch-Partenkirchen		Bilbao
Italy	Rome		Toledo
	Casaccia	U.S.S.R.	Moscow
	Milan		Lahemaa
	Venice	Portugal	Jeronimo Monastery
Netherlands	Vlaardingen	Canada	Dorset
	Eibergen	U.S.A.	Research Triangle
	Vredepeel		Steubenville
	Wijnandsrade		

8 The economic cost of material damage to buildings and building materials

SUMMARY

Estimating the financial benefits from reductions in ambient pollutant concentrations involves many steps, each of which has associated uncertainties. Currently the greatest of these are associated with the damage functions that are available. If damage functions are used to estimate the effect of pollutants on materials, then the most successful approach is to apply advanced sensitivity analysis to a decision analysis model, backed by sound economic judgement. An alternative approach is to compare expenditure on repair and replacement in 'clean' and 'polluted' areas. Any estimates that are made are minimum valuations because some benefits, such as aesthetic quality, cannot be quantified.

8.1 INTRODUCTION

Policies on the reduction of acid deposition should ideally be linked to the relationship between pollutants, environment and resultant damage. The cost effectiveness and benefits of policies require careful study. Since relationships between the reduction of pollutant emissions and their benefit cannot be established directly, intermediate stages are needed, namely determination of atmospheric concentrations, physical material damage, total building stock involved, and the cost of damage.

8.2 CONTEXT OF THE CHAPTER

In general terms, there are three elements to the problem of assessing the cost of acid deposition damage to buildings:

(i) Determining the stock of buildings or building materials/components exposed from inventory (see Chapter 3);

(ii) Determining the exposure and the damage due to acid deposition (see Chapters 2, 6);

(iii) Determining the costs of this damage, or conversely, the benefits due to reduced acid deposition.

This chapter is entirely concerned with the last element. Nevertheless, it must be pointed out that a key problem for the whole report is element (ii) above. The mechanisms of material and building damage due to acid deposition and especially the distinction between damage caused or enhanced by acid deposition and that due to 'normal' weathering, ageing and decay, are not well understood. Furthermore, it is not always easy to see or assess damage (lay people visiting monuments may not perceive it or interpret it correctly) and there are no agreed means of measuring damage either physical or aesthetic. This means that the attribution of damage to acid deposition, and hence of cost savings or benefits arising from reduced pollution, is technically difficult and uncertain. One has to rely on what is reasonably acceptable.

8.3 COMPLEXITY OF DETERMINING DAMAGE COSTS

The external envelope of a building is made up of many different elements (walls, roofs, guttering and downpipes, window framing, windows, doors, lintels, steps, etc) each of which may be made of different materials, and may be protected in a number of ways. For example, brick walls may be left uncovered or can be painted or rendered while window joinery may be of timber, plastic or metal which may or may not be painted, etc.

Each of these elements, depending upon the material of construction and type of protection, will require maintenance, repair, or replacement if the fabric of the building is to be retained. Painted surfaces need repainting to prevent weather penetrating to the materials beneath. Mortar in brick walls will need repointing to maintain the integrity of the structure. Metal cladding and zinc flashing will need replacing as its protective coating is lost and deterioration occurs.

Therefore over the life time of the building there will be a series of maintenance, repair and replacement costs. The extent that reduced acid deposition will lead to a reduction in the rate of deterioration of building materials should be reflected in the frequency with which they have to be maintained, repaired or replaced. Hence, reduced pollution levels should lead to reduced average annual repair and maintenance costs.

The early approaches to the assessment of the costs of SO_2 damage (eg OECD 1981), and some of the later inventory approaches, concentrated upon determining the quantity of materials exposed to acid deposition and calculating a new material replacement cost. However, such an approach is inadequate because the costs to society of deterioration relate not only to the material that is affected, but also to the costs of maintenance, repair and replacement of the item or building component that is affected. The service life, maintenance cycle, rate of deterioration and the costs of repair, etc are as much a function of the building component itself as they are of the material of which it is made or with which it is protected.

For example, the surface of painted brickwork has an average life of some 10 years and costs £3.27 per square metre to repaint (1987 prices). Wooden joinery, however, normally needs repainting every 3-6 years at a cost of £2.18 per square metre. In addition, the costs of repair and maintenance of the painted component vary even more. Painted brick walls require re-pointing (on average every 40 years at a cost of £12.60 per square metre), whilst timber joinery requires replacing (even with good maintenance) every 25-30 years at a cost of £880 per square metre. The extent to which reduced pollution levels extend maintenance cycles and replacement periods affects the savings that are made (Williams 1985).

From this example, it is clear that an inventory of the stock of buildings at risk is required which gives not only the materials but also the amount of each building component in terms of material and any protective coating that may be present. Methods developed by ECOTEC (Williams 1985) aim to provide this type of inventory.

In order to identify the key materials and components for consideration, an assessment has to be made of the types of building materials and their susceptibility to acid deposition. This can then be related to the range of building components used in the building fabric to identify those which should be included in the inventory. Table 3.1 illustrated the relationships between materials and building components.

8.4 AN ASSESSMENT OF THE COSTS OF DAMAGE AND SAVINGS FROM ABATEMENT IN SO₂

This section gives examples of calculated benefits, the theory of which is given in more detail later in the chapter. The basic assumption is made that the costs of maintenance of the building stock are a minimum valuation of the benefits received (or saving achieved) by the building owner/occupant. The word 'minimum' is used because any consumer surplus (ie a benefit valued by the purchaser as higher than the purchase price) is ignored, as are other benefits such as the aesthetic pleasure that a passer-by or resident gets in a well maintained area. Hence, the cost of repair and maintenance is viewed as a minimum assessment of the value of maintaining the building in good condition.

It is possible to calculate the impact on maintenance and repair cycles and service life of reducing ambient sulphur levels. In order to do this, information is required on the average length of life of building components. Their required maintenance, repair, replacement cycles and practices need to be established, as do the costs of materials and labour which can be derived from standard cost guides and price books. There are, however a number of problems in performing such calculations. These primarily relate to the limitations of current damage functions and direct or synergistic effects of pollutants on materials.

The limited number of suitable damage functions, particularly with respect to stone, is the weakest link in assessing the savings to be made from abatement in pollution concentrations. However, estimates of the value of savings resulting from the control of sulphur emissions have been made (Salmon 1970; PAU 1972; OECD 1981). The OECD (1981) study of Western Europe estimated an annual per capita saving of £2.70, and the application of the methods described in Chapter 3 lead to an estimate of £2.50 per capita for Birmingham (based on a 30% reduction in SO_2 emissions).

In the Birmingham study, the major items to provide cost savings in a reduced pollution regime were galvanised steel (roof and cladding), walls and slate roofs. In an historic town such as Lincoln, the items are similar but also include rendering, plaster work and stone elevations. Predicted cost savings to be made from reduced air pollution in the Conservation Area in Lincoln are slightly higher than those to be made in the normal building stock in the city. This reflects not so much the use of more vulnerable materials, but rather, it reflects the higher cost of workmanship required while local materials are used both within and without the conservation area, certain materials in the older buildings, eg medieval timber framed buildings, are not protected but are still not susceptible to damage by acid deposition.

It should be noted that the cost savings, or benefits, are minimum estimates since they do not include aesthetic qualities. Neither do the assessments include individual buildings of national importance (eg cathedrals, castles, stately homes).

8.5   COST OF DAMAGE: ACCURACY OF DATA

Some studies (PAU 1972) have related costs of damage directly to pollutant concentrations by comparing expenditure on repair and replacement in 'clean' and 'polluted' areas, but the more common approach is to use a 'damage function' of the sort described in the previous chapter. The costs are then derived from:

(i)   The damage calculated for each material studied;

(ii)   The unit cost of repair or replacement of materials;

(iii)   The amount of materials exposed.

Uncertainty is introduced at each stage and this must be taken into account when deciding on methods to assess costs. For example it would not be logical to conduct a very accurate inventory of materials if the accuracy of damage estimates for the materials varies between ±100%. Once the weakest link has been identified and if its accuracy lies outside permissible pre-defined limits, then this link must be changed with re-evaluation of the others as necessary.

The previous chapter showed that damage functions are not available for all economically important materials. One study in the United States (Salmon 1970) ranked materials in order of the total economic damage they sustain (Table 8.1). The table serves as a useful indicator of the relative importance of materials in the United States. The economic loss is obviously subject to uncertainty and conditions in the United Kingdom today will certainly be different from those in the study.

Table 8.1   Ranking of materials economic losses in the United States due to atmospheric pollution (Salmon 1970)

MATERIAL	ECONOMIC LOSS ($ Million 1970)
Paint	1195
Zinc	778
Fibres	358
Cement and Concrete	316
Nickel	260
Rubber	194
Tin	144
Plastics	126
Aluminium	114
Copper	110
Carbon steel	53.3
Building brick	24.2
Paper	22.1
Leather	20.6
Wood	17.6
Building stone	17.6
Brass and Bronze	13.4
Magnesium	13.0
Others	23.2

The total uncertainty introduced in going from unit damage to unit cost cannot be quantified in the same way as uncertainties in either the estimate of unit damage or the estimate of total damage from unit damage and quantity of material at risk.

8.6   TREATMENT OF UNCERTAINTY

Assessment methods differ in the ease with which they can deal with this uncertainty. Uncertainty can be dealt with in a number of ways:

(i)   No treatment;

(ii)   Verbal acknowledgement;

(iii)   Simple sensitivity analysis;

(iv)   Advanced sensitivity analysis;

(v)   Value of information analysis.

The complete omission of any treatment of uncertainty (8.6 (i)) is usually only found in partial analyses dealing with limited aspects of the problem. Any analysis aiming at completeness is forced to acknowledge the existence of uncertainty, verbally at least (8.6 (ii)). Unfortunately, a verbal caveat can easily become detached from a set of figures as these pass through the decision-making hierarchy.

A simple sensitivity analysis (8.6 (iii)) will

normally report the effect of introducing some standard variation, often plus or minus a few per cent, in each input to the analysis. This can certainly give a feel for the importance of each input, but little more, since there is nothing to indicate whether the standard variation is a reasonable approximation to the true degree of uncertainty surrounding this input. Sensitivity analysis is still too easy to detach from the base results.

An advanced sensitivity analysis (8.6 (iv)) includes the treatment of uncertainty from the outset, rather than as an afterthought. A typical example would include each input variable as a probability distribution, with a Monte Carlo or other advanced sampling scheme to produce a probability distribution of the outputs (Mackay *et al* 1979). The importance of the uncertainty in each input can be specifically represented, by stating the correlation between inputs and outputs for instance. This information can inform any research effort of the best directions in which to concentrate their resources. There is little danger of the uncertainties in the results being overlooked by accident and further protection can be provided by coding the outputs in one of a scheme of notations designed to guard against deliberate misrepresentation (Funtowicz & Ravetz 1986).

If a decision analysis technique has been used to calculate the results then the possibility exists of taking the sensitivity analysis one stage further by explicitly calculating the value of better information (8.6 (v)) about any of the inputs. One drawback is that the calculated value applies only to the single decision under consideration. This may seriously undervalue the information if it also has value in a different sphere, eg deciding which buildings should be listed. The calculations can also become quite involved. Microcomputer acid rain models can take many hours to find the value of information. For this reason, advanced sensitivity analysis of a decision analysis model backed by sound economic judgement will often be the best compromise.

8.7 VALUATION METHODS

There is a large literature concerned with the theory of evaluation particularly as it relates to non-market goods and a large number of methodologies for determining values. Only a few of these methodologies have, to date, been employed in attempting to assess the benefits of reduced acid deposition damage to buildings and building materials.

Before describing the various methodologies, it is useful to divide the building stock into three classes:

(i) The general stock of buildings that have no architectural or historic classification;

(ii) Buildings in conservation areas where it is the general area and type of development that is of historical or architectural interest;

(iii) Individual buildings (listed grade I, churches, cathedrals, etc) that are of specific architectural and/or historic merit.

Methodologies for dealing with the general building stock (repair costs, physical damage function and materials loss approaches, optimal materials use approaches, and differential property value methodologies) are the subject of work that has been carried out in the context of a study for the Commission of European Communities (CEC) (Williams 1985). To the extent that group 8.7 (ii) above may be subject to methodologies appropriate for individual historic buildings, they are included under this category.

Individual historic monuments and buildings (group 8.7 (iii)) are of such cultural importance that they require special evaluation approaches to assess their worth. In addition the approach to defining damages for this group (eg aesthetic impact) is different since functional aspects are less relevant.

8.8 EVALUATING BENEFITS AND COSTS

To assess any proposed strategy for pollution control it is necessary to weigh up both advantages and drawbacks. The advantages will usually consist of reduced damage in a variety of areas, such as crops and building materials. The drawbacks will consist mainly of the costs of the specific pollution control measures. The weighing up process will involve a method of valuing the reduction in damage, so that it can be compared to the cost of obtaining it. Given the divided nature of the evidence, a way of dealing with the uncertainties involved will also be required.

The range of methods proposed for weighing up benefits against costs is shown below:

(i) No valuation

(ii) Paretian cost-benefit analysis

(iii) Decision-making cost-benefit analysis

(iv) Decision making

The 'No valuation' option implies that the benefits are expressed solely in physical units, such as percentage drop in crop yield or weight of material eroded. This may be an attractive option to the analysts involved, given the difficulties encountered by any attempt to do more. It deals with the problem of valuation by leaving it to the unaided intuition of the reader of the report. It has been defended by pointing to the political nature of many of the valuations required, but there is ample evidence to suggest that even for political decisions some further form of analysis is beneficial.

Paretian cost-benefit analysis can give this help by attempting to express a market value for each benefit. For example, an increased yield of crops would be valued at the market price for each extra tonne produced. This technique has been developed by economists over many years and text books abound, such as Mishan (1982). Even if the price cannot be observed directly, economists have developed techniques involving indirect deduction, such as the use of property values, to estimate the financial benefit of noise reduction. However, if there is no market at all in which the benefit appears, there is no way in which a valuation according to independent economic principles can be made. Unfortunately this appears to be true for many of the most important benefits from pollution control. The value people put on the knowledge that historic buildings are protected even though they have no intention of visiting them, is one example. Analyses using the Paretian form of cost-benefit analysis are unlikely to give such intangible benefits their full weight and may omit them entirely.

In the decision-making form of cost-benefit analysis the valuation of these intangible benefits by the decision maker is accepted as a legitimate input to the analysis (Sugden & Williams 1978). This takes the analyst far from the original aim of cost-benefit analysis, the non-controversial totting up of financial advantages and drawbacks. The incorporation of subjective inputs does not

seem to fit with the claimed objectivity of cost-benefit analysis. Yet the results of the analysis still have to be expressed in the economist's medium of money, which is least suited to handle just those parts of the acid deposition problem that make it controversial eg the rights of unborn generations, the concept of the ecosystem as a fragile web of interactions, etc.

If decision aids are required, it may make sense to enlist the techniques that have been developed specifically to serve this end, such as multi-attribute decision analysis which has sprung from the interaction of psychology with engineering. Decision analysis is based on the subjectivity of all important inputs to the acid deposition problem and has developed suitable methods to handle them correctly (Koeney & Raiffa 1976). The drawback is the need to give up the image of objectivity provided, however spuriously, by cost-benefit analysis.

8.9 CONCLUSIONS

Estimating the financial benefits from reductions in ambient pollutant concentrations involves many steps, each of which has associated uncertainties. Currently the greatest of these are associated with the damage functions that are available. If damage functions are used to estimate the effect of pollutants on materials, then the most successful approach will be to apply advanced sensitivity analysis to a decision analysis model, backed by sound economic judgement. An alternative approach is to compare expenditure on repair and replacement in 'clean' and 'polluted' areas. Any estimates that are made are minimum valuations because benefits, such as aesthetic quality, cannot be quantified. Nor is the value of nationally important buildings taken into account.

9 Protection and preservation of building materials

SUMMARY

Traditional materials and practices in the construction industries have actively developed over a period when atmospheric pollution was much more intense than it is now. In most cases, reasonably durable materials and economical building methods have evolved. Demands on building materials are not entirely utilitarian, they include elements of aesthetics and of fashion, so that the treatments specified are not always the economically optimal ones.

The literature on treatments for the preservation and protection of metals suggests that the most serious deficiency is in information on contemporary corrosion rates, how they have altered with falling SO_2 emissions and whether all metals have benefited equally. Further information is also required on the synergistic effect of NO_x and SO_2.

Many of the non-metallic building materials considered (eg brickwork) appear to perform adequately for most purposes and do not require protective treatments to improve their resistance to industrial atmospheres. The outstanding exceptions to this generalisation are limestone and other calcareous stones, especially in heritage buildings. These need to be preserved for long periods without damage to fine detail although they tend to be built from susceptible materials. Many aspects of the mechanisms of damage are still not understood and as will be pointed out, no really satisfactory protective or remedial treatments are currently available. Further research on the microstructure of stone should provide a more reliable foundation for the development of future protective treatments

9.1 INTRODUCTION

Most, if not all, building materials will be damaged to some extent by weathering even if there were no industrial pollution. All rain in the United Kingdom contains some chloride and sulphate derived from seawater aerosol and small amounts of sulphur dioxide from natural sources. These inevitably give rise to some deleterious effect on materials. There are also effects caused by fluctuations in temperature and humidity, oxidation due to ozone and ultraviolet irradiation, and biologically induced degradation. However, the preceeding chapters of this report have demonstrated that many deleterious effects are accelerated by anthropogenic pollutants. Because of this damage, appreciable costs can be incurred relating to protective and preservative measures, maintenance and replacement.

This chapter is concerned with treatments used to improve the durability of materials under atmospheric attack, either in new components ('protection') or for reconditioning or prolonging the life of components that have already been damaged by weathering ('preservation'). An attempt will be made to specify the extent to which various materials are at risk in the British climate and what can be done to prolong life and repair damage.

The range of materials involved is very large and degradation processes may be complex. Certain building materials show such poor inherent resistance to weathering that they can only be considered for use in United Kingdom conditions because satisfactory protective schemes are available. Others are traditionally expected to endure for long periods with no applied protection and only occasional maintenance or cleaning. Protective finishes such as paints or plastics coatings or tiles are often used for decorative purposes, or to modify surface texture or to facilitate cleaning and may therefore need to be considered as materials in their own right. In most cases, however, the substrate and the coating must be considered together. Degradation of the coating will expose the substrate to atmospheric attack. This in turn will cause further damage to the coating through mechanical disruption or chemical degradation. Both types of process may be accelerated by atmospheric pollutants.

Materials can undergo many deleterious effects (Yocum *et al* 1982), ranging from discolouration or loss of gloss on paints to crumbling of building stones or catastrophic detachment of claddings through the corrosion of fasteners. The conditions of exposure are also complex and may not correspond to those envisaged by the designer. Consequently degradation may result from details of design or workmanship as much as from selection of materials (Schmid 1984; Jacobson & Lindgren 1971; Ashton & Sereda 1982).

It is not possible to cover points of detail, which are treated at length in design manuals and codes of practice, but it is worth noting that many problems can be avoided at the design stage, particularly by eliminating incompatible combinations of materials, moisture traps, etc.

An attempt has been made to indicate the techniques available, particularly for the restoration of building components found to be damaged after a period of service. However, no estimates of costs have been included since these vary so much according to the circumstances of each case.

References to appropriate literature are given wherever possible, but in some cases information has had to be based on discussion with building industry specialists.

9.2 METALS

Metals are used in building for structural strength, in the form of sheeting, as formed parts such as screws and other fasteners, channel sections and pipes and also for decorative features. The properties of most interest are strength and formability and the cheapest materials, such as plain carbon steels, are used in many applications where their susceptibility to corrosion makes some degree of protection vitally important.

The problems presented by metal corrosion are well understood. Effective and reasonable economic protective measures are available for almost all of them so long as they are dealt with at the design stage and not left to be handled by maintenance procedures after trouble has developed.

9.2.1 Iron and Steel

Plain carbon steels differ from the other materials discussed in that their resistance to corrosion in open exposure to the atmosphere is so poor that it is impracticable to use them without protection except in very large sections such as railway lines. Rates of corrosion vary from 3 μm year^{-1} to at least 1,000 μm year^{-1} according to wetness and atmospheric contamination. In present-day British conditions, rates will be below 100 μm year^{-1} in the absence of special factors. Even this can present serious problems due to perforation of thin sections, failure of fasteners, seizure of screw-threads, leakage of components by the expansion of corrosion products, discoloration etc. By no means are all building components exposed to the weather and unprotected mild steel may well be used in suitably enclosed locations, but there is usually some risk of unforeseen entry of water by leakage or condensation.

Low-alloy steels, containing perhaps 0.3-0.5% copper with no minor additions of other metals, corrode in industrial atmospheres at 20-30% of the rate of plain steels and form relatively stable rust layers whose appearance is often considered attractive. They are sometimes used without protection in relatively large sections. They are not, however, seen at their best in the British climate and they appear to lose much of their advantage over mild steel in relatively clean air.

Stainless steels (containing 12-18% chromium, usually with some nickel) are much more resistant to rusting and higher quality steels are almost completely unaffected by industrial atmospheres. They are used for a wide variety of purposes for cladding and fasteners and for high-quality architectural ironmongery. With proper selection of material for the prevailing conditions, they require no protection.

(a) Protection

The protection and maintenance of structural steel is discussed in publications by the British Standards Institution (1977). The commonest form of protection for steelwork is undoubtedly painting. Paints provide versatile and economical solutions in most situations and in buildings where reasonable freedom from mechanical damage can be assumed. Paints do not protect merely by excluding the oxygen and water needed for the corrosion process but also their high electrical resistance and their ability to exclude strong electrolytes from the metal surface are often more important factors. Many paints contain substances that act as corrosion inhibitors and this gives a limited tolerance to both mechanical damage and the presence of corrosive agents. It is, however, necessary to clean corrosion products from metal surfaces before painting since they usually contain salts that stimulate further corrosion and may in turn damage the paint film. This requirement for good surface preparation adds considerably to the cost of painting and is to a large extent attributable to atmospheric corrosion. (Figure 9.1).

Once a paint film has begun to fail (often at previously contaminated spots, gaps left by faulty application, or mechanical damage), hygroscopic corrosion products usually form under the film and produce water-filled blisters which detach the paint from the metal and also contribute to corrosion by prolonging wetting.

Estimates have been made of the total cost of painting in the United Kingdom (PAU 1972; Vernon 1957; DTI 1971; DOE 1972), and these have been quoted in estimating the total cost of corrosion. However, it must be remembered that much painting is carried out for reasons other than corrosion protection and that repainting is often done for aesthetic reasons before it is strictly necessary. Nevertheless, repainting intervals are probably reduced by atmospheric pollution even if the degradation is not allowed to progress to the point where the substrate is damaged. General information on painting can be found in publications of the British Standards Institution (1977; 1982a) and Hamburg and Morgans (1979).

Paint films may fail through several different mechanisms and these may act in combination. Direct attack by atmospheric agents on the paint vehicle or the pigments can lead to cracking, blistering, or generalised thinning of the film itself. Degradation of the substrate (which may produce bulky products) can damage or detach the film by either mechanical or chemical action. Atmospheric pollutants can play a part in all these mechanisms (Yocum et al 1982); but attack on the substrate is stimulated by mechanical defects caused by rough handling, poor preparation, or uneven application. There is evidence that some pigments, particularly extenders based on calcium carbonate, are attacked by atmospheric SO_2 (Campbell et al 1973; Holbrow 1962; Kaiser 1977) and that certain paint vehicles may be hydrolysed by

highly acidic solutions (Holbrow 1962; Funke et al 1984). The drying of certain paints may also be impeded by atmospheric SO_2 (Holbrow 1962; Funke et al 1984). Besides these effects there is little other evidence that SO_2 plays a major part in damage mechanisms associated with paint films and in general degradation effects appear to be more closely related to ultra-violet radiation and ozone (Hamburg & Morgans 1979; Kucera 1976; Gaynes 1985). Kulis (1979) has shown that the life of paint films was considerably reduced by the heavily polluted atmosphere of industrial areas in Czechoslovakia but no pollution data are quoted.

It must be emphasised that a great many different polymers and pigments are used in paint formulations. They differ greatly in chemical reactivity and rely on several different mechanisms for producing a stable dry film. Some paints rely on simple evaporation of solvent, some on the coagulation of particles from a water dispersion, some on oxidative cross-linking of unsaturated esters, some on complex reactions between partially formed polymer intermediates to produce a cross-linked final product, etc. A vast range of chemical types is involved with differing reactivities to aggressive agents. Physical properties also vary considerably, corresponding to requirements for differing substrates and conditions of exposure. Exterior paints must accommodate dimensional changes due to changes in humidity and temperature without cracking or creeping or becoming detached from the substrate.

Apart from paint, iron and steel are protected by a number of other coating systems including factory-applied ('coil-coated') polymeric materials, metal coatings and ceramics. Factory coatings are often sufficiently robust to withstand forming operations and can include zinc under-layers which provide protection at gaps and cut edges. They represent an attractive possibility for light-gauge components such as roofing, side walling etc, which does not depend on the vagaries of preparation and painting on site. Factory coatings are generally similar in their properties to paints, but less information is available about their performance. The polymers used and their modes of application vary considerably. There is little evidence that these materials, as a class, are more severely degraded on industrial areas than rural areas. However, as with paints, the consequences of poor application or mechanical damage will be exacerbated in polluted atmospheres.

A variety of metal coatings are used for decoration and/or protection. The coatings may consist of metals which are in themselves highly resistant to corrosion, such as chromium or nickel, or may act by forming a resistant layer of corrosion product. The latter type will be discussed in the next section. Chromium is extremely resistant to atmospheric attack, but is difficult to fabricate and is usually applied as thin, highly stressed, electro-deposits. Since these are invariably porous they are likely to be undermined and ultimately destroyed by the corrosion of the underlying materials. This process is, of course, accelerated by atmospheric pollutants. Sodium chloride derived from seawater aerosol or from road salts probably plays a more important part in attacking chromium-plated steel than SO_2. There is a large literature on the testing of electroplated articles, but it is difficult to relate the results of tests under artificial conditions, which are normally only used for quality assurance, to atmospheric corrosion. There are few specific reports on the effects of individual pollutants other then SO_2, but in general the degradation is associated with corrosion of the substrate at gaps. Nickel is well known to be attacked by atmospheric SO_2 and it should therefore be considered along with zinc and cadmium. However, in mild conditions, rates of corrosion are sufficiently slow for nickel plate to be used for decorative purposes.

Ceramic coatings in the form of vitreous enamel are extensively used for decorative items and cladding. While the coating remains intact they provide excellent protection to the underlying material. The coatings consist essentially of highly resistant borosilicate glasses and no reports have been found of damage specifically identified with atmospheric pollutants.

(b) Preservation

The rusting of iron and steel in contaminated atmospheres produces deposits of corrosive salts and thorough cleaning is therefore essential before any restorative treatment such as repainting of rusted metal is undertaken. This requires removal of the whole of the corrosion product which is often present in deep pits and crevices. Since treatments such as wire-brushing are quite inadequate the use of abrasives, grit-blasting, or acid pickling is essential if the components have become heavily rusted. These treatments are time-consuming and expensive and it may not be possible to treat all the corroded surfaces. It is therefore advisable to repaint before the components have become heavily corroded. Equally, it is desirable to protect new components to prevent rusting during construction.

Preservative treatments should also be applied before damage has become too severe. It is often more economic to replace material than to use restorative treatments because the treatments are

frequently undermined by corrosion products that could not be removed.

9.2.2 Zinc and Galvanizing

Zinc is a comparatively reactive metal. It will corrode in contact with iron by setting up an electrochemical cell which maintains a potential difference that protects the iron from attack but corrodes the zinc. This 'sacrificial' action is useful because it protects exposed areas of steel on galvanizing components, eg damaged areas and cut edges. Corrosion of zinc also forms a coherent layer of sparingly soluble basic salts. These tend to protect the underlying metal from further attack. Corrosion products on zinc, unlike those on iron, are not recycled to cause further corrosion.

In rural areas zinc sheet or galvanized steel may give satisfactory life without further treatment. In industrial or marine environments, however, zinc coatings are usually employed as a foundation for painting, or galvanized steel may be exposed unprotected until the coating thickness is reduced to the minimum safe value and then painted. Zinc coatings have the particular advantage of providing a margin of safety if the paint film is damaged.

(a) Zinc Coatings

Zinc is used in the form of sheet and solid sections. It is also applied as coating to steel by a number of techniques, including hot-dipping, flame-spraying, electrodeposition and cementation. For fasteners used in sheltered conditions, thin coatings are adequate and the anti-corrosive action of these can be improved by chromate treatments.

Fasteners for severe conditions (eg wall-ties) are heavily hot-dipped. Zinc can also be used as a pigment in priming paints. This gives many of the advantages of zinc coatings with considerable ease and flexibility of application.

The behaviour of zinc coatings on steel is very similar to that of zinc itself and the useful life of a coating is closely related to the initial thickness. Costs of galvanizing, and of eventually painting galvanised components, represent an important part of the total cost of atmospheric corrosion. Apart from painting ungalvanized steel, this was in fact the only item for corrosion included in the OECD (1972) calculation on costs of pollution.

Recommendations on the application and usage of zinc coatings are contained in British Standards Institution publications (1960; 1971).

Cadmium is used for corrosion protection in much the same way as zinc. Its high cost confines its use to thin electrodeposits particularly for electronic equipment, for screws and light-gauge fasteners. Little has been reported on the effects of atmospheric pollutants.

Aluminium coatings are also used for protection of structural steel and for some high-temperature applications. Its properties are similar to those of zinc except that its action depends much more on the imperviousness of the oxide film and less on sacrificial protection.

(b) Protection and Preservation

In many cases it is considered more cost-effective to allow galvanised steel to approach failure and then replace it rather than apply extra protection. This applies particularly to light-gauge, relatively inexpensive items (such as wire fencing) in relatively clean conditions. The painting of zinc and galvanised steel has often presented difficulties because of the tendency of the paint vehicle to react with zinc to produce an intermediate water-sensitive layer that impairs adhesion. Consequently zinc has often been allowed to weather to form sufficient corrosion product to provide a key for painting. Satisfactory primers are available for unweathered zinc and their use is preferable to reliance on weathering. Proper storage of zinc-coated items on site is important.

For severely-rusted exposed steelwork, grit-blasting followed by metal-spraying is the obvious restorative treatment, but in many cases this will not be practicable. In some applications zinc-rich priming paints provide a satisfactory technique for restoration.

9.2.3 Copper

Copper and its alloys are used for roofing, water supply, statuary, and other architectural metalwork. The resistance of copper to atmospheric corrosion is considerably better than that of zinc and more than an order of magnitude better than that of steel. Corroded surfaces are protected by a layer of sparingly soluble hydrated salts containing sulphate, chloride and, in some cases, carbonate. Copper roofing sheet has given satisfactory service for 200 years in town atmospheres. Consequently copper alloys are seldom painted for protection. However, aesthetic damage can occur to brick or stone when the presence of hydrated copper salts causes staining.

In special conditions, particularly in contact with packaging materials, copper alloys can suffer deep and unsightly penetration due to attack by organic acids ('bronze disease') and a special electrochemical technique has been devised for removing the corrosion product from crevices.

(a) Protection

Protection is not usually applied to copper alloys because many people prefer the appearance of the corrosion product. The original bright appearance of copper can be maintained by applying an acrylic lacquer containing an effective corrosion inhibitor, usually benzotriazole.

(b) Preservation

Since copper and its alloys form stable corrosion products, surface preservation techniques are only required when 'bronze disease' or other gross outbreaks of corrosion occur. Preservation treatments usually demand similar techniques to those used by restorers. Treatment is more staightforward than for iron or steel as it is easier to neutralise deposits of corrosion products.

9.2.4 Lead

Lead has good resistance to atmospheric corrosion in industrial areas because its sulphate is extremely insoluble. Throughout historic times lead has been used for roofing and flashings, water-supply, and architectural metalwork. The most frequently reported cause of failure is contamination with organic acids in water supplies or under organic debris. Lead can be painted, but this is not usually considered necessary for protection.

9.2.5 Aluminium

Aluminium and its alloys are highly resistant to atmospheric attack. They are used in the form of sheet for cladding and flashings and occasionally in statuary and decorative items. Metallic aluminium is highly unstable in the thermodynamic sense, but is protected by an extremely coherent and unreactive oxide film. This can, however, be broken down in highly acidic or alkaline conditions, or in solutions of chlorides. Atmospheric corrosion is somewhat accelerated in industrial areas, but the rate is still more than an order or magnitude slower than that in steel and the attack slows down considerably over time. There is a tendency to pitting, especially in industrial areas. This is accentuated where the corrosion products are not washed by rain. Little quantitative information is available on the effects of gaseous pollutants other than SO_2. Very rapid corrosion can occur in contact with salts of copper or mercury, but these do not usually constitute a problem in atmospheric exposure. Problems are, however, sometimes caused by contact with fixings or timber preservatives.

Anodising, in which the natural oxide film is thickened by electrochemical treatment and can be dyed or treated with corrosion inhibitors, provides a range of finishes which enhance appearance and are resistant to corrosion. Aluminium can be successfully painted, but generally this is not considered necessary except for copper-containing alloys and thin sections of cladding.

9.3 NON-METALLIC MATERIALS

The building industry employs a large range of non-metallic materials. Many of these may react directly with the atmosphere, while others, unreactive in themselves, may suffer degradation due to thermal cycling, frost damage and salt crystallisation. Atmospheric pollutants may effect many of these materials in complex ways. Few quantitive measurements have been carried out on the degradation of most buildings and still fewer systematic investigations of damage mechanisms.

9.3.1 Wood

In a typical building a considerable area of wood is exposed to the environment and a good deal more is liable to occasional exposure. However, the effect of atmospheric pollution is very small compared to the problems caused by water and humidity.

Woods are complex materials and they differ greatly in their porosity and water resistance. Only a few hardwoods are suitable for outdoor exposure without protection. The great bulk of exposed wood consists of softwood which if left unprotected would become waterlogged and develop excessive porosity through bacterial action, or would rot (Banks & Evans 1984).

(a) Protection

A large variety of different protective schemes and fungicide treatments are available for wood,

including paints, oils, waxes and resin-impregnation. Paints can improve water-resistance, but to do so effectively they must resist mechanical damage and cracking caused by changes in temperature or water content. There is little evidence that atmospheric pollutants, such as SO_2 and NO_x, affect paint films to any great extent at the lower SO_2 contents that now prevail in towns in the United Kingdom. General advice on painting buildings is given a publication by the British Standards Institution (1982a).

(b) Preservation

Wood that has split can be treated with adhesive polymer-based fillers before repainting. Similar materials can be used with reasonable success to fill rotted sections of timber after removing the affected wood and applying suitable preservatives.

9.3.2 Stone

Serious attack on natural stone by atmospheric pollutants is confined to limestones and calcareous sandstones. Other types of stone, particularly granite, are little affected by pollution. Stone damage problems also occur through frost damage or by crystallisation of soluble salts. The latter may appear independently of industrial pollutants, in sea-spray, soil-water, or by contact with contaminated building materials such as mortar (Amoroso & Fassina 1983; Koller 1979; BRE 1984). Details of mechanisms of decay and the role of individual pollutants are described in Chapter 6.

(a) Protection

No really satisfactory means are available for long term protection of stone from atmospheric attack. The use of paint on stonework may lead to problems with flaking and crumbling similar to those it is intended to avoid. Although some contractors advocate the use of silicone water repellents to delay resoiling of cleaned stone, this is not generally recommended (BRE 1983; British Standards Institution 1982b).

(b) Preservation

There is no cheap method of preserving large areas of stonework. The most common practice is to dress off loose stone and repoint (for surface damage) or, alternatively, to completely replace grossly damaged stones. The cost of preservation can be justified only for buildings and art-objects of particular interest. Individual objects are sometimes removed from the building for display in a controlled environment and a replica may be placed in the original position. Apart from these options conservators have the choice of silane-based preservatives or techniques based on lime.

Silane-based preservatives can be used on limestones or sandstones. They are intended to tie the friable outer stone on to the sound inner core, to prevent damaging crystallisation cycles and, in the case of limestones, protect against acid pollutants. One such silane system is Brethane (Price 1981) developed by the Building Research Establishment.Other silane systems are commercially available but their application and penetration cannot be so closely controlled as the catalytic reaction of Brethane. Trials carried out by BRE over recent years have shown them to be less effective (Price 1981). However, alternative silane-based systems are being introduced and these may be suitable for use on large scale areas of stonework.

Lime-based conservation techniques are suitable for use on limestones only. The object is to replace lost calcium carbonate by impregnating the porous stone with lime which is then carbonated by atmospheric carbon dioxide. In summary, the lime method involves cleaning the stone with a lime poultice, consolidation with up to 40 applications of limewater, repairs being carried out in lime mortar, and finally the application of a protective lime-based shelter coat (Ashurst 1984).

Lime-based techniques are sometimes favoured by conservators because they involve only natural materials. In contrast, silane treatment introduces 'foreign' chemicals which, once applied, cannot be removed. In practice it would be difficult to remove *in situ* any treatment applied to stone and the trend now is to require that the stone will be retreatable at a subsequent date.

Lime treatment has been tested in the laboratory (Price & Ross 1984) but no explanation of the reasons for its success was found. It may be due to the care and attention lavished on the stone throughout the lengthy treatment.

Work continues on estimating the success of lime-based and silane treatments in preserving stone and protecting it from further attack by acid pollutants. The life expectancy of the lime treatment is controlled by the original composition of the shelter coating and the exposure conditions. After the coating breaks down it can simply be replaced. Exposure trials of stone treated with Brethane have shown that consolidation of samples is still good after

10 years. Brethane treated stone also shows strong resistance to attack by dilute sulphuric acid. (Figure 9.2.)

9.3.3 Brick, Cement and Mortar

There is little evidence of damage to brickwork by industrial pollutants. Bricks in themselves are highly stable and therefore require no special protection or preservation. Most damage is by frost action.

9.3.4 Roofing Tiles

So far as is known, atmospheric pollutants do not contribute significantly to attack on clay roofing tiles. Generally tiles degrade through frost damage, crystallisation of salts, or biological action (eg lichens, algae, etc). Concrete tiles may show superficial erosion (see Section 9.3.7).

9.3.5 Natural Slate

Those slates which contain carbonates among their mineral constituents are susceptible to the same types of atmospheric attack as similar building stones. There is no record of silane or lime-based conservation techniques being used to preserve or protect slates against acid pollutants. The BSI recommend that slates which pass all the tests (including acid immersion) in BS680: Part 2: 1971 are used in towns. However, the testing criteria for slates have been the subject of some contention over the years because certain slates of proven durability have failed to meet all the test requirements.

9.3.6 Renderings

Acid pollutants are not known to adversely affect renderings. However, even a good quality render will fail if the substrate is subject to sulphate attack.

9.3.7 Concrete

The main forms of degradation of concrete are carbonation of the concrete structure, surface erosion, and corrosion of steel reinforcement.

(a) Preservation and Protection

Erosion of concrete (up to half aggregate dimension) is a surface effect, and it does not normally affect the structure.

Anti-carbonation treatments applied to concrete include limewash, two pack polyurethane and epoxy systems, bitumen, acrylics, and chlorinated rubbers. However, organic coatings are susceptible to ultra-violet light and to movement of the substrate. Protective or preservative measures against corrosion of reinforcements include the use of stainless steel or galvanised reinforcements, organic coatings for reinforcements, adding corrosion inhibitors to the concrete mix, and the use of coatings to prevent carbonation and entry of salt (Engelfield 1977). Information on corrosion of steel reinforcements in concrete is contained in reports by Muller-Golchert (1982), BRE Digest (1982), and the Concrete Society (1984).

9.3.8 Glass

There is no evidence that modern glass is affected by atmospheric pollution except by soiling and by surface leaching. The soiling is easily removed by regular washing and the leaching increases durability (see section 6.3.4). Medieval glass is considerably less durable than modern glass because its chemical composition leads to a weakening of its network structure (Cox *et al* 1979).

(a) Preservation and Protection

The confusion over the mechanisms which cause glass to decay makes it difficult to develop preservation or protection methods. It has been suggested that medieval glass should be protected from both water and SO_2 by coating. However, as water is thought to be aggressive there is concern that water vapour would penetrate organic coatings and that the coatings could react with the glass. Exposure studies of coatings are taking place in York at the present time (1987). Until these studies are completed the best policy appears to be to install externally ventilated and externally mounted glazing to protect important window glass.

9.3.9 Polymers

Although polymers are not used for load-bearing elements in buildings, a variety of materials are used for structural elements such as glazing bars, as well as for sheeting, for rainwater goods, and for sealing strip. Other important applications are in mastics and sealing compounds. Most of the structural components are of PVC or glass-reinforced polyester, but large quantities of nylon and polypropylene extrusions and injection-

mouldings are also used. Sealing materials use a variety of rubbers (natural and synthetic), and also bitumens.

The most important reactions in the degradation of many polymers are photo-oxidation and oxidation by ozone (Davies & Sims 1983; Grassie 1977-1984). These reactions may produce a surface layer of brittle, cross-linked material, cracks initiating in this layer may propagate into unaffected polymer. There may also be photo-oxidative effects on filler particles and reinforcing fibres at the surface. Additives have been developed which control most of these problems. Little systematic work appears to have been carried out to define the contribution of industrial pollutants to these deleterious effects. No reports have suggested that either structural materials or sealing and jointing compounds deteriorate more rapidly in industrial atmospheres than in rural areas, probably because of the general incorporation of anti-oxidants in polymer formulations.

9.4 CONCLUSIONS

Traditional materials and practices in the construction industries have actively developed over a period when atmospheric pollution was much more intense than it is now. In most cases reasonably durable materials and economical building methods have evolved. Where the inherent resistance of materials is inadequate, effective treatments have been developed for protection and maintenance. Demands on building materials are not entirely utilitarian, they include elements of aesthetics and of fashion, so that the treatments specified are not always the economically optimal ones. The general subject of materials development is beyond the scope of this review, but little evidence has been found that technical problems associated with present levels of pollution play an important part in determining the direction of development.

9.4.1 Metals

The literature on treatments for the preservation and protection of metals suggests that the most serious deficiency is in information on contemporary corrosion rates, how they have altered with falling SO_2 emissions, and whether all metals have benefited equally. Further information is also required on the synergistic effect of NO_x and SO_2. Given these uncertainties, it is difficult to derive a national strategy for protection of materials.

Most of these uncertainties relating to the effect of present day atmospheric pollution should be clarified through the National Material Exposure Programme (see Section 7.5). More work is however needed on validating, or otherwise, synergistic effects, which must include laboratory tests as well as exposure trials.

9.4.2 Non-Metals

Many of the non-metallic building materials considered (eg brickwork) appear to perform adequately for most purposes and do not require protective treatments to improve their resistance to polluted atmospheres. The useful lives of most buildings are brought to an end by changes in use before major renovations become necessary through degradation of materials. The outstanding exceptions to this generalisation are limestone and other calcareous stones, especially in heritage buildings. These need to be preserved for long periods without damage to fine detail although they tend to be built from susceptible materials. As a result of past and present industrial pollution, these have undoubtedly suffered, and are probably still suffering, damage both to structural members and to the detail of art-objects. Many aspects of the mechanisms of damage are still not understood and as has been pointed out, no really satisfactory protective or remedial treatments are currently available. Therefore further research on both the mechanisms of attack and the effect of atmospheric conditions on rates of decay is necessary to define the problem. Such research is very important in connection with remedial treatments.

Stone is a porous material and much of the damage results from crystallisation of salts in the pore structure. It is not clear at present to what extent soluble salts, once formed, continue to pass into the pores of the stone. Consequently protective coatings applied after surface cleaning may still be undermined by continued reaction deep inside the stone (memory effect). A mechanism of this type certainly operates on rusted iron and steel where hygroscopic product left in deep pores produced by corrosion can initiate fresh corrosion under paint films. Further research on the microstructure of stone should provide a more reliable foundation for the development of future treatments. Work on the detailed characterisation of the samples from the NMEP will provide a starting point for further research on this important topic. Development of cleaning processes to ensure removal of corrosive reaction products, with minimal damage to the remaining material, also appears to be an important objective.

10 Conclusions and Recommendations

10.1 CONCLUSIONS

10.1.1 The evidence put forward in this report shows all of the features to be expected if atmospheric pollution is accelerating weathering and damage. However, the evidence is not yet unequivocal that present rates of weathering on historic buildings are significantly different from those in the past. The evidence for metals shows a reduction in the rate of weathering for steel, but little to this effect for other metals. Nor does the present available evidence indicate if 'safe' levels exist or which atmospheric components are the major causes of damage – moisture, frost, salts, SO_2, particulates, NO_x, etc. As in other areas of the environment, damage is probably due to a combination of stress factors. (Chapter 2; Chapter 6).

10.1.2 The pollutants involved are generally SO_2, particulates, oxides of nitrogen (NO_x), carbon dioxide and, for organic materials, ozone (O_3). In coastal areas, chloride plays an important part. (Section 2.1; 6.4).

10.1.3 The rates of weathering (estimated from weight losses) for stone samples measured by exposure at Garston, Westminster and St Paul's Cathedral appear to fall within the ranges of 'natural' rates. However, measurements made on historical building components are very much higher than 'natural' rates and show a variation with aspect and position. (Section 2.2; 2.3).

10.1.4 Higher rates of weathering of new stone were measured at urban sites compared to rural sites in the study of stone in South-East England, and in a NATO/CCMS study. However, in the latter, rates in coastal regions were as high as in urban areas. A similar rural/urban increase has been recorded for iron, copper and tin. (Sections 2.3.1(b), (c); 2.3.3).

10.1.5 Chemical changes in building materials were observed on new stone samples exposed in South-East England. Sulphate concentrations increased on both sheltered and exposed samples. Nitrate and chloride concentrations also increased but less dramatically. No direct correlation was observed between atmospheric SO_2 content and weight loss. The presence of sulphates on metals (particularly iron and zinc) is also indicative of the role of pollutants in corrosion. (Sections 2.3.1(c); 2.3.3).

10.1.6 In view of the pressing need for further information on current rates of decays of building materials, BERG has recommended the establishment of a national monitoring network to investigate the decay of new building materials in the United Kingdom. The United Kingdom Government has already accepted this recommendation and the National Materials Exposure Programme (NMEP) came into operation in April 1987. The network consists of 29 sites at which samples of different building materials (stone, steel, etc) will be exposed for a minimum of four years. Information on meteorological conditions and atmospheric pollution will be collected from all sites during this period. Four of the national sites are included in an international network that is being operated concurrently for the UNECE. (The BRE has co-ordinated the programme and will be responsible for collation and analysis of the NMEP data.) (Section 7.5).

10.1.7 Annual average concentrations of SO_2 in urban atmospheres in the United Kingdom have decreased significantly in the last 30 years. Although historical data beyond the early decades of the 20th century are scarce, it is likely that prior to the 1950s, concentrations of SO_2 in Central London had changed little since the 17th century. Since the 1950s and 1960s, urban concentrations of SO_2 have decreased at a faster rate than total emissions of SO_2, reflecting the proportional shift in emissions from low and medium level urban sources to high level emitters often located outside urban areas. (Section 4.2).

10.1.8 Annual average urban concentrations of black smoke have decreased by a factor of ten during the last 30 years. An important development has been the changes in the relative contributions from the major sources, most notably the shift from domestic solid fuel to vehicle emissions. (Section 4.3).

10.1.9 Over the past 50 years emissions of NO_x in the United Kingdom have roughly doubled, primarily because of the increase in road traffic. However, there has been no apparent trend in urban air concentrations of NO_x or ozone at the relatively few sites where measurements have been made over the past ten years or so. (Section 4.4).

10.1.10 Elevated concentrations of CO_2 known to occur in urban areas, may affect rainfall pH and also be an important factor in the degradation of limestone and concrete but further studies are required in these areas. (Section 4.5; Section 6.3.1a).

10.1.11 Chloride concentrations in rainwater are generally dominated by marine sources. In industrial areas, however, other sources may be important. No definite figures for chloride emissions exist for the United Kingdom but coal combustion appears to be the most important man-made source. Emissions and ambient chloride arising from coal combustion are likely to have decreased markedly over the past 30 years or so, particularly in urban areas. (Section 4.7).

10.1.12 The atmospheric dispersion models discussed in this report are capable of providing accurate estimates of the contributions to airborne pollutant concentrations from the major source categories. However, the calculations presented in this report used very approximate estimates of emissions in the towns studied. In order to provide improved information on the urban contribution to pollutant concentrations around historic buildings and other areas, detailed emission inventories are needed. Such information is essential to the construction and evaluation of pollution control policies. (Chapter 5).

10.1.13 In an investigation of a sample of 26 towns in the United Kingdom the external contribution to total annual (1983) mean SO_2 concentrations for the town was estimated to be at least 40% and in most cases 50% or greater. The exceptions were those towns with a population of several hundred thousand which lay outside the higher SO_2 regions of the United Kingdom (eg those broadly outside the 20-30 $\mu g\ m^{-3}$ annual mean), where the external contribution was 30-35%. A larger proportion of

the external contribution arises from United Kingdom power stations. The amount varies from 20-30% of the external contribution in remote areas, up to 60-70% in regions downwind of the major East Midlands and Yorkshire power stations. (Section 5.3.1).

10.1.14 For episodic conditions in urban areas, calculations demonstrate that low and medium level sources tend to be more important than high level sources, but under suitable meteorological conditions high level sources may dominate for short periods. (Sections 5.3.2; 5.3.3).

10.1.15 It seems likely that in urban areas of the United Kingdom the turbulent transfer of pollutants, particularly SO_2, by dry deposition is the most important mechanism for transferring pollutants to the surfaces of buildings and materials. However very few measurements of wet or dry deposition or of deposition via fogs and mists have been carried out in urban areas. Furthermore, significant gaps exist in our understanding of turbulent transfer in built-up environments. Deposition velocities in urban areas are at present very uncertain, severely limiting the accuracy with which calculations of deposition of pollutants to building and material surfaces can be made. Moreover, there are uncertainties in the nature of the variation of deposition velocities with wind speed and wetness of the surface. These can further complicate estimates of source attribution. (Section 6.2).

10.1.16 The interactions between building materials and pollutants are very complex, and many variables are involved. Deposition of pollutants onto building surfaces (as particulates, aerosols, or gases) depends not only on their atmospheric concentrations and the turbulence profile at the boundary layer, but on the strength and direction of the wind and the rain intensity, both in the open and immediately around the deposition site (micro-climate). Once pollutants are present on the material surface, interactions will vary with the position of the material (eg its degree of exposure to wind and rain). The time-of-wetness of the surface is particularly important. Finally, the nature, structure and reactivity of different materials, and the degree to which they are protected, will determine the degree of damage caused. (Sections 3.3; 6.2.5).

10.1.17 The complex character of deposition and damage is further compounded by the large number of uncertainties, particularly in reaction mechanisms and the role of different pollutants. Natural processes also play a significant role in the degradation of building materials, particularly frost, wind and rain. Atmospheric carbon dioxide is thought to lead to dissolution of calcareous stone but its importance relative to other pollutants is not quantified. However, it is clear that calcareous stones (limestone, marbles, and calcareous sandstones), iron, steel and zinc, and non-durable medieval glass are at greater risk than other building materials. It is also clear that water, Cl^-, and SO_2 (and associated salts) are all important in accelerating damage relative to 'natural levels'. Some degradation of stone is probably caused by recrystallisation of salts derived from pollutants that were deposited in the past (the 'memory effect') but its effect relative to degradation initiated by present pollutants is not known. Other pollutants (such as NO_x and particulates) are probably also important, but at present the evidence against them is less conclusive. (Sections 3.2; 6.3).

10.1.18 Laboratory studies have clearly shown that atmospheric pollutants can damage most building materials. Pollutants were thought to act independently but there is now increasing evidence that they may act synergistically. It is also clear that moisture (relative humidity, time of wetness, drying cycles, etc) is critical in the degradation of materials. (Section 2.1; 6.4).

10.1.19 Estimating the number of buildings and the amount of material at risk from acid deposition is a difficult and complex process. However, it is essential for any estimate of the cost of damage and consequently for any form of cost benefit analysis. (Section 3.5).

10.1.20 The costs of damage to building materials by atmospheric pollution not only varies with the material type, but also depends upon the nature of the component and its function within the building. Using building components as the central focus for assessing damage and damage costs allows aspects of service life and maintenance cycles to be incorporated into the damage cost assessment process. (Section 3.5).

10.1.21 Cost assessments, based on the Probability-Distribution approach, are applicable to all buildings, including the general stock of historic buildings in conservation areas and produce a minimum estimate of the willingness to pay to reduce damage. Historic buildings and monuments of national and international importance require a quite different evaluation approach since their cultural and aesthetic significance may be much greater than their replacement value. (Section 3.5).

10.1.22 Currently available damage functions do not seem to perform well in predictive use and in many cases are unsatisfactory. The main reason

for this is that damage functions do not take account of the complex relationships of environmental variables (eg chloride, NO_x/SO_2 synergism, rain intensity). One approach is to apply a damage function relating overall rates of corrosion to pollutant levels, and possibly a wetness parameter, to area averages for the relevant pollutants obtained from deposition models. The results can then be applied to an estimate of the stock-at-risk for various materials obtained from survey evidence or from scaling up on the basis of models (Chapter 7).

10.1.23 Estimating the financial benefits from reductions in ambient pollutant concentrations involves many steps, each of which has associated uncertainties. Currently the greatest of these are associated with the damage functions that are available. If damage functions are used to estimate the effect of pollutants on materials, then the most successful approach is to apply advanced sensitivity analysis to a decision analysis model, backed by sound economic judgement. An alternative approach is to compare expenditure on repair and replacement in 'clean' and 'polluted' areas.

Any estimates that are made are minimum valuations because benefits, such as aesthetic quality, cannot be quantified. Nor is the value of nationally important buildings taken into account. (Chapter 8).

10.1.24 Traditional materials and practices in the construction industries have actively developed over a period when atmospheric pollution was much more intense than it is now. In most cases reasonably durable materials and economical building methods have evolved. Where the inherent resistance of materials is inadequate, effective treatments have been developed for protection and maintenance. Demands on building materials are not entirely utilitarian, they include elements of aesthetics or fashion, so that the treatments specified are not always the most economic. (Chapter 9).

10.1.25 Many of the non-metallic building materials considered (eg brickwork) appear to perform adequately for most purposes and do not require protective treatments to improve their resistance to polluted atmospheres. The useful lives of most buildings are brought to an end by changes in use before major renovations become necessary through degradation of materials. The outstanding exceptions to this generalisation are limestone and other calcareous stones, especially in heritage buildings which need to be preserved for long periods without damage to fine detail although they tend to have been built from susceptible materials. (Chapter 9).

10.2 RECOMMENDATIONS AND FUTURE WORK

As a result of reviewing the available evidence relating to the effect of acid deposition on buildings and building materials in the United Kingdom the group has listed ongoing research (Appendix E) and identified a number of areas which require further study, and has made the following recommendations:

10.2.1 There should be a long-term commitment, both financially and of research capabilities, to co-ordinated studies related to the effect of acid deposition on building materials.

Such studies will allow a realistic assessment of current rates of degradation compared to historic rates and the extent of that degradation. It will also help to establish if rates are altered by lower levels of some atmospheric pollutants.

The National and International Materials Exposure Programmes, the first steps towards these aims, have already been initiated. However, the quality of data available from these studies would be improved if continuous pollutant monitoring was available on more sites.

10.2.2 Present emissions of pollutants need to be established if accurate dispersion models are to be constructed and applied to towns and cities of interest.

In order to assess the potential benefits of future emission reductions on buildings and materials damage, atmospheric models need to be used. However, it is essential that detailed emission inventories are developed for the towns and cities of most interest in order that assessments of acceptable accuracy can be made.

10.2.3 A better understanding of the relative importance of 'episodic' (acute) and long-term exposure to pollutants, and evidence on possible 'threshold' levels is required.

These studies should combine laboratory and site programmes to obtain evidence for which pollutants are most damaging, how important synergistic reactions are, and the relationship between pollutant levels and damage. In particular these studies should include synergistic effects of SO_2, NO_x, and particulates, the effect of NO_x acting alone, and the effect of relative humidity. All future studies should use realistic levels of pollutants.

10.2.4 The relative importance in urban areas of wet and dry deposition, the mechanisms of deposition and associated variables (particularly deposition velocity) should receive further attention.

In order to improve understanding of the damage processes the mechanisms by which the deposited pollutants react with the materials, particularly the effect of electrolyte film composition on metal corrosion and the effect of rainwater composition (eg pH) on stone degradation, should be assessed. Environmental variables (eg relative humidity, surface wetness) and micro-climate should also be included in the overall assessments.

10.2.5 It is necessary to develop methods of predicting damage at both site specific, regional and national scales.

In order to improve assessments of damage improved definition of damage functions, particularly for non-metallic materials is required. This will allow application of closely co-ordinated laboratory and site studies, development of new methods to characterise damage and degradation, and consideration of the problems of relating continuously monitored pollution data to intermittent degradation data.

10.2.6 The protection and preservation of materials, particularly stone and non-durable medieval glass, requires further research.

The role of salts in stone after protection is not clearly understood, nor are the structural relationships between preservation or protection techniques and the materials being treated fully understood. The development of new processes for cleaning and removing corrosive contaminants, whilst causing minimal damage would also be beneficial. Further studies are required on the protection of medieval glass including the use of coatings and secondary glazing.

10.2.7 There is a need to extend work on the integrated costs of damage by pollution to individual buildings.

There is a need to look in more detail at the issues of costs associated with reliability, engineering applications, the actual costs of replacement or improved specification of building components, etc. This is fundamental to getting a proper estimate of the cost of materials damage and could be best carried out by making a number of case-studies of individual buildings in a different pollution environments. There are certain building categories, eg city centre office blocks, which would particularly benefit from improved and more detailed analysis.

10.2.8 **The 'inventory approach' to stock-at-risk needs to be calibrated and extended over a range of large areas or regions to assess the degree to which secondary source data and 'grossing up' can be used.**

The inventory approach using 'identikits' has been shown to be a useful technique but it needs to be calibrated where necessary and extended to include a full range of building materials and architectural styles used in the United Kingdom. An assessment is also required of the degrees to which this inventory approach and secondary source data (eg Ordnance Survey Maps) can be used to obtain data on the stock of materials at risk in the United Kingdom.

10.2.9 **It is suggested that for a small number of key buildings, surveys are made to determine the financial benefits which could result from preservation and protection.**

Given that currently very little is known about the relative size of the value placed upon the benefits of preservation and protection to historic buildings and monuments of national significance, it seems important to determine some indication of this value to provide a general perspective to cost savings figures.

10.2.10 **BERG should continue to meet and a further report should be produced following the completion of the first phase of the NMEP in 1991.**

The group does not consider that the evidence examined to date is adequate to enable its terms of references to be fully met. It draws attention to further data which are likely to become available and especially to the National Materials Exposure Programme which will provide information on rates of weathering for a range of typical building and construction materials exposed at 29 sites throughout the United Kingdom by 1991. The group recommends that information arising from the NMEP should be kept under continuous review a second BERG report should be produced at this time.

References

Ailor, W.H. (1969). Aluminium corrosion at urban and industrial locations, *J.Pr.ASCE* (Struct.Div.), 2141-2160.

Ailor, W.H. (1974). *ASTM STP*, 558, 75.

Ailor, W.H. (Ed) (1982). *Atmospheric Corrosion*. New York: John Wiley and Sons.

Alderborn, J. (1971). Investigations of weathered glass surfaces with the scanning microscope. *OECD report on scientific research on glass* (Ref: DAS/SPR/71.35), 244-254.

Al-Ismail, M.A. (1981). Effects of ammonium chloride particulate matter on aluminium. M.Sc thesis, University of Manchester Institute of Science and Technology (Unpublished).

Ambler, H.R. (1960). Atmospheric salinity and metal corrosion at various places in Great Britain, *J.Appl.Chem.*, 10, 213-225.

Amoroso, G.G. and Fassina, V. (1983). Stone decay and conservation: Atmospheric pollution, cleaning, consolidation, and protection. *Materials Science Monographs* 11. Oxford: Elsevier.

Apling, A.J., Keddie, A.W.C., Weatherley, M-L.P.M. and Williams, M.L. (1977). The high pollution episode in London, December 1975. *Warren Spring Laboratory Report* LR 263 (AP).

Arshadi, M.A., Johnson, J.B. and Wood, G.C. (1983). The influence of an Isobutane – SO_2 pollutant system on the earlier stages of the atmospheric corrosion of metals. *Corros.Sci.*, 23, 763-776.

Ashton, H.E. and Sereda, P.J. (1982). Environment, microenvironment and the durability of building materials. *Durability of Building Materials*, 1, 49-65.

Ashurst, J. (1984). The cleaning and treatment of limestone by the 'lime method', Part 1. *Monumentum*, 233-252.

Atkinson, T.C. and Smith, D.I. (1976). The erosion of limestone. In *The Science of Speiliology* (Ford, T.D. and Cullingford, C. Eds). 151-177. Academic Press.

Atmospheric Dispersion Modelling Working Group (ADMWG) (1979). A model for short and medium range dispersion of radionuclides released to the atmosphere. *National Radiological Protection Board Report* NRPB-R91.

Atteraas, L. and Haagenrud, S. (1982). Atmospheric corrosion testing in Norway. In *Atmospheric Corrosion* (Ailor, W.H. Ed). 873-892. New York: John Wiley and Sons.

Ball, D.J. and Hume, R. (1983). The relative importance of vehicular and domestic emissions of dark smoke in Greater London in the mid 1970s, the significance of smoke shade measurements and an explanation of the relationship of smoke shade to gravimetric measurements of particulates. *Atmos.Environ.*, 17, 169-181.

Ball, D.J. and Radcliffe, S.W. (1979). An inventory of sulphur dioxide emissions to London's air. *Greater London Council Research Report* 23. London: GLC.

Banks, W.B. and Evans, P.D. (1984). Degradation of wood surfaces by water. *Proc. 18th Annual Meeting, International Research Group on Wood Preservation, Stockholm.*

Barker, C.D. (1978). A comparison of Gaussian and diffusivity models of atmospheric dispersion. *Central Electricity Research Laboratory Report* RD/B/N4405.

Barry, P.J. (1975). Stochastic properties of atmospheric diffusivity. *Atomic Energy of Canada Ltd Report* AECL-5012.

Barton, K. (1973). *Protection Against Atmospheric Corrosion*. New York: John Wiley and Sons.

Barton, K. and Czerny, M. (1980). The relationship between the properties of the medium and the kinematics of atmospheric corrosion of steel, zinc, copper and aluminium. Assessment of results of first five-year stage in corrosion testing program of the member-nations of the COMECON. *Zash.Metallov.*, 16, 387-395 (Russian); *Prot. of Met.*, 16, 301-308.

Bawden, R.J. (1985). The deterioration of natural building stones: a review. *Central Electricity Research Laboratories Report* TPRD/L/2821/N85.

Benarie, M. and Lipfert, F.L. (1986). A general corrosion function in terms of atmospheric pollutant concentrations and rain pH. *Atmos.Environ.*, 20, 1947-1958.

Bennett, M., Rogers, C. and Sutton, S. (1986). Mobile measurements of winter SO_2 levels in London 1983-1984. *Atmos.Environ.*, 20, 461-470.

Bettelheim, J. and Littler, A. (1979). Historical trends of sulphur oxide emissions in Europe since 1850. *Central Electricity Research Laboratory Report* PL-GS/E/1/79.

BISRA (1938). Fifth report of the corrosion committee. *Iron and Steel Special Report* 21.

BISRA (1959). Sixth report of the corrosion committee. *Iron and Steel Special Report* 66.

Blaeuer, C. (1985). Weathering of Bernese sandstones. In the proceedings of the Vth *International Congress on Deterioration and Conservation of Stone* (Lausanne), 381-390.

Booth, F.F. and Goddard, H.P. (1965). The corrosion behaviour of aluminium alloys in marine atmospheres. *Congres Internationale de la Corrosion Marine*, 313-331.

Brimblecombe, P. (1977). London air pollution 1500-1900. *Atmos.Environ.*, 11, 1157-1162.

Brimblecombe, P. (1986). *Air – Composition and Chemistry.* Cambridge: Cambridge University Press.

British Standards Institution (BSI) (1960). Electroplated coatings of cadmium and zinc on iron and steel. *BS 1706:1960.*

British Standards Institution (BSI) (1971). Hot dip galvanised coatings on iron and steel articles. *BS 729:1971.*

British Standards Institution (BSI) (1977). Code of practice for protective coating of iron and steel structures. *BS 5493:1977.*

British Standards Institution (BSI) (1982a). Code of practice for painting buildings. *BS 6150:1982.*

British Standards Institution (BSI) (1982b). Code of practice for cleaning and surface repair of buildings. Part 1: Natural stone, cast stone and clay, and calcium silicate brick masonry. *BS 6270:1982.*

Britton, S.C. (1976). Tin versus corrosion. *ITRI Pub.* No.510.

Britton, S.C. and Angeles, R.F. (1951). Weathering tests of tin – zinc alloy coatings on steel. *Metallurgia*, 44, 185-191.

Buck, D.M. (1913). Copper in steel – The influence on corrosion. *J.Ind.Eng.Chem.*, 5, 447-452.

Building Research Establishment (BRE) (1971). Sulphate attack on brickwork. *BRE Digest* 89. London: HMSO.

Building Research Establishment (BRE) (1983). The selection of natural building stone. *BRE Digest* 269. London: HMSO.

Building Research Establishment (BRE) (1982). The durability of steel in concrete. *BRE Digest* 263-265 (3 parts). London: HMSO.

Building Research Establishment (BRE) (1983). Cleaning external surfaces of buildings. *BRE Digest* 280. London: HMSO.

Building Research Establishment (BRE) (1984). Decay and conservation of stone masonry. *BRE Digest* 177. London: HMSO.

Burgess, C.F. and Aston, J. (1913). Influence of various elements on the corrodibility of iron. *J.Ind.Eng.Chem.*, 5, 458-462.

Butlin, R.N., Cooke, R.U., Jaynes, S.M. and Sharp, A.D. (1985). Research on limestone decay in the United Kingdom. In the proceedings of the Vth *International Congress on Deterioration and Conservation of Stone*, (Lausanne), 537-546.

Campbell, G.G., Schurr, G.G. and Slawikowski, D.E. (1973). Assessing air pollution damage to coatings. *Am.Chem.Soc.Div. of Organic Coatings and Plastic Chem.*, 33, 99-117.

Carter, V.E. (1968). Atmospheric corrosion of aluminium and its alloys: Results of six-year exposure tests. *ASTM STP*, 435, 257-270.

Carter, V.E. (1982). Atmospheric corrosion of non-ferrous metals. In *Corrosion Processes* (Parkins, R.N., Ed). 77-113.

Chamberlain, A.C., Garland, J.A. and Wells, A.C. (1984). Transport of gases and particles to surfaces with widely spaced roughness elements. *Boundary Layer Meteorology*, 29, 343-360.

Chan, W.H. and Chung, D.H.S. (1986). Regional-scale precipitation scavenging of SO_2, SO_4, NO_3 and HNO_3. *Atmos.Environ.*, 20, 1397-1402.

Chandler, K.A. and Kilcullen, M.B. (1968). Survey of corrosion and atmospheric pollution in and around Sheffield. *Brit.Corros.J.*, 3, 80-84.

Charola, E.A. (1987). Acid rain effects on stone monuments. *J.Chem.Education*, 64, 436-437.

Ciarallo, A., Festa, L., Piccioli, C. and Raniello, M. (1985). Microflora action in the decay of stone monuments. In the proceedings of the Vth *International Congress on Deterioration and Conservation of Stone*, (Lausanne), 607-616.

Clarke, J.F. (1969). A meteorological analysis of carbon dioxide concentrations measured at a rural location. *Atmos.Environ.*, 3, 375-383.

Clarke, S.G. and Bradshaw, W.N. (1953). Tests of the protective value of metallic coatings under sheltered conditions (marine atmosphere). *J.Appl.Chem.*, 3, 147-154.

Cofer, W.R.III, Schryer, D.R. and Rogowski, R.S. (1980). The enhanced oxidation of SO_2 by NO_2 on carbon particulates. *Atmos.Environ.*, 14, 571-575.

Cofer, W.R.III, Schryer, D.R. and Rogowski, R.S. (1981). The oxidation of SO_2 by NO_2 on carbon particulates. *Atmos.Environ.*, 15, 1281-1286.

Concrete Society (1984). Concrete reinforcement corrosion. *Concrete Society Technical Report* 26.

Corbel, J. (1959). Erosion et terrain calcaire: vitesse d'erosion et morphologie. *Ann.Geogr.*, 68, 97-120.

Cox, G.A. and Pollard, A.M. (1977). X-ray fluorescence analysis of ancient glass: the importance of sample preparation. *Archaeometry*, 19, 45-54.

Cox, G.A., Heavens, O.S., Newton, R.G. and Pollard, A.M. (1979). A study of the weathering behaviour of medieval window glass from York Minster. *J.Glass Studies*, 21, 54-75.

Davies, A. and Sims, D. (1983). *Weathering of Polymers*. Barking: Elsevier Applied Science.

Del Monte, M., Sabbioni, C. and Vittori, O. (1981). Airborne carbon particles and marble deterioration. *Atmos.Environ.*, 15, 654-652.

Del Monte, M., Sabbioni, C. and Vittori, O. (1984). *Sci.Total Environ.*, 36, 369.

Department of Environment (DOE) (1972). Report of the committee on building maintenance. London: HMSO.

Department of Environment (DOE) (1984). Acid rain: The Government's reply to the fourth report from the Environment Committee. London: HMSO.

Department of Environment (DOE) Digest (1978-1986). *Digest of environmental protection and water statistics*. London: HMSO.

Department of Trade and Industry (DTI) (1971). Report of the committee on corrosion and protection. London: HMSO.

Derwent, R.G. (1986). Atmospheric ozone and its precursors. *ETSU Report* R-38.

Drew, D.P. (1975). The limestone hydrology in the Mendip Hills. In *Limestone and Caves of the Mendip Hills* (Smith, D.I. and Drew, D.P. Eds). 171-213. Newton Abbot: David and Charles.

Eckhardt, F.E.W. (1985). Mechanisms of the microbial degradation of minerals in sandstone monuments, medieval frescoes, and plaster. In the proceedings of the Vth *International Congress on Deterioration and Conservation of Stone* (Lausanne), 643-652.

Edney, E.O., Stiles, D.C., Spence, J.W., Haynie, F.H. and Wilson, W.E. (1986). Laboratory investigations of the impact of dry deposition of SO_2 and wet deposition of acidic species on the atmospheric corrosion of galvanized steel. *Atmos.Environ.*, 20, 541-548.

Englefield, R. (1977). Carbonation of concrete: its significance and control by coatings. *Defrazet*, 31, 353-359.

Engelfried, R. and Toller, A. (1985). Effect of moisture and SO_2 content of air on carbonation of concrete. *Betonwerk u. Festigteil-Technik*, 11, 722-729.

Evans, T.G. (1969). Atmospheric corrosion behaviour of stainless steels and nickel alloys. *Proc.4thInt.Conf.Met.Corr.*, NACE, Houston, 408-417.

Evans, U.R. (1960). *The Corrosion and Oxidation of Metals*. London: Arnold.

Fink, F.W., Buttner, F.H. and Boyd, W.K. (1971). Technical-economic evaluation of air pollution corrosion costs on metals in the United States. *Batelle Memorial Institute, Columbus Laboratory, Ohio Report* PB-198-453.

Fitz, S. (1985). Do air pollutants damage historic stained-glass windows? *Electrochemical Society Extended Abstracts*, 85-2, 202.

Franey, J.P. and Graedel, T.E. (1985). Corrosive effects of mixtures of pollutants *J.Air pollution Control Assoc.*, 35, 644-648.

Frenzel, G. (1985). The restoration of medieval stained glass. *Scientific American*, 252(5), 100-106.

Friend, J.N. (1929a). The relative corrodibilities of ferrous and non-ferrous metals and alloys. *J.Inst.Met.*, 42, 149-155.

Friend, J.N. (1929b). Final report on the relative corrodibilities of various commercial forms of iron and steel. Results of prolonged exposure to the atmosphere, and general conclusions from all the tests. *J.Iron Steel Inst.*, 61 (Carnegie Scholarship Memoirs, 18), 61.

Funke, W., Haagen, H., Krieg, E. and Zeh, J. (1984). *Proc. XVII FATIPEC Congress, Lugano*, 4, 177-187.

Funtowicz, S.O. and Ravetz, J.R. (1986). *J.O.pl.Res.Soc.*, 37, 243-247.

Garland, J.A. (1977). The dry deposition of sulphur dioxide to land and water surfaces. *Proc.Roy.Soc.Lond.*, A354, 245-268.

Garland, J.A. (1978). Dry and wet removal of sulphur from the atmosphere. *Atmos.Environ.*, 12, 349-362.

Gaynes, N.I. (1985). Ozone is the culprit. *Metal Finishing*, 83, 45-47.

Gilbert, P.T. (1953). The effect of impurities in the metal on the rate of corrosion of zinc and galvanised coating in the atmosphere. *J.Appl.Chem.*, 3, 174-181.

Grassie, N. (Ed) (1977-1984). *Developments in Polymer Degradation.* Barking: Elsevier Applied Science.

Guidobaldi, F. (1981). Acid rain and corrosion of marble. An experimental approach, the conservation of stone II. *Preprints of the contributions to the International Symposium, Bologna.* 483-497.

Guidobaldi, F. and Mecchi, A.M. (1985). Corrosion of marble by rain. The influence of surface roughness, rain intensity and additional washing. In the proceedings of the Vth *International Congress on the Deterioration and Conservation of Stone* (Lausanne). 467-474.

Guttman, H. and Sereda, P.J. (1968). Measurement of the atmospheric factors affecting the corrosion of metals. *ASTM STP*, 435, 326-359.

Haagenrud, S., Kucera, V. and Gullman, J. (1982). Atmospheric corrosion testing with electrolytic cells in Norway and Sweden. In *Atmospheric Corrosion* (Ailor, W.H. Ed), 669. New York: John Wiley and Sons.

Hache, A. (1962). Influences des microclimats sur les essais de corrosion a l'atmosphere en milieu tempere et tropical. *Corrosion et Anticorrosion*, 10, 68-77.

Hailstone, S.C. (1969). Cast aluminium alloys in a marine environment for twenty years. *Metallurgia*, 316, 136-140.

Hakkarainen and Ylasaari (1982). Atmospheric corrosion testing in Finland. In *Atmospheric Corrosion* (Ailor, W.H. Ed). 787-794. New York: John Wiley and Sons.

Hamburg, H.R. and Morgans, W.M. (1979). *Hess's Paint Film Defects,* 3rd edition. London: Chapman and Hall.

Harrison, W.H. (1981). The conditions for sulphate attack on brickwork. *Chemistry and Industry*, 18, 636-639.

Haynie, F. (1980). Theoretical air pollution and climate effects on materials confirmed by zinc corrosion data. In *Durability of Building Materials and Components* (Sereda, P.J. and Litvon, G.G. Eds), 157-175. American Society for Testing and Materials.

Haynie, F.H. and Upham, J.B. (1971). Effects of atmospheric pollutants on corrosion behaviour of steels. *Mat.Perform.*, 10, 18-21.

Haynie, F.H., Spence, J.W. and Upham, J.B. (1976). Effects of gaseous pollutants on materials – a chamber study. *Environmental Protection Agency (USA) Report* EPA-600/3-76-015.

Higgins, R.I. (1956). *J.Res.Brit.Cast Iron Res.Assoc.*, 6, 165.

Holbrow, G.L. (1962). Atmospheric pollution: Its measurement and some effects on paint. *J.Oil Colour Chem.Assoc.*, 45, 701-718.

Honeyborne, D.B. (1982). *The building limestones of France.* London: HMSO.

Honeyborne, D.B. and Harris, P.B. (1958). Discussion on the structure of porous building stone and its relation to weathering behaviour. In *The Structure and Properties of Porous Materials* (Everett, D.H. and Stone, F.S. Eds), *Proc. of the Tenth Symposium of the Colston Research Society*, 363

House of Commons Environment Committee (1984). Acid rain. *Fourth Report, Session 1983-1984.* London: HMSO.

House of Commons Environment Committee (1986). Follow-up to the Environment Committee report on acid rain. *First Special Report, Session 1985-1986.* London: HMSO.

Hudson, J.C. (1929). Atmospheric corrosion of metals – Third (experimental) report to the corrosion research committee (British non-ferrous metals research association). *Trans.Farad.Soc.*, 25, 177-252.

Hudson, J.C. (1930). The effects of two years atmospheric exposure on the breaking load of hard-drawn non-ferrous wires. *J.Inst.Met.*, 44, 409-431.

Hudson, J.C. (1935). The effect of five years atmospheric exposure on the breaking load and the electrical resistance of non-ferous wires. *J.Inst.Met.*, 56, 91-102.

Hudson, J.C. and Banfield, T.A. (1946). The protection of iron and steel by metallic coatings. Results of five years exposure tests. *J.Iron and Steel Inst.*, 154, 229-264.

Hudson, J.C. and Banfield, T.A. (1947). Discussion on Hudson and Banfield 1946. *J.Iron and Steel Inst.*, 157, 349-368.

Hudson, J.C. and Stanners, J.F. (1953). Protection of iron and steel by metallic and non-metallic coatings II. *J.Iron and Steel Inst.*, 175, 381-390.

Hudson, J.C. and Stanners, J.F. (1953b). The effect of climate and atmospheric pollution on corrosion. *J.Appl.Chem.*, 3, 86-96.

Hutchins, J.D. and McKenzie, M. (1973). Characterisation of bridge locations by corrosion and environmental measurements. First year's results. *Transport and Road Research Laboratory Report* LR550.

Jacobson, L. (1972). Effect of air pollutant precipitation on building surfaces. *National Swedish Building Research Summaries Report R23.*

Jacobson, L. and Lindgren, H. (1971). Impact of air pollution and climate on buildings: deterioration of exterior surfaces. *Chalmers University of Technology, Goteburg Report.*

Jaynes, S.M. (1985). Studies of building stone weathering in South-East England. *Ph.D Thesis University of London* (Unpublished).

Jaynes, S.M. and Cooke, R.U. (1987). Stone weathering in southeast England. *Atmos.Environ.*, 21, 1601-1622.

Johansson, L.G. (1985). A laboratory study of the influence of NO_2 and SO_2 on the atmospheric corrosion of steel, copper, zinc and aluminium. *Electrochemical Society Extended Abstracts*, 82(2), 221-222.

Johansson, L.G., Lindqvist, O. and Mangio, R. (1986). A study of the corrosion of calcareous stones exposed to humid air containing SO_2 and NO_2. In preprinted papers of the *Air and Pollution Symposium, Rome, October 15-17 1986.*

Johnson, J.B., Skerry, B.S. and Wood, G.C. (1983). Influence of smoke and isobutane on corrosion in sulphur dioxide polluted environments. *J.Electrochem.Soc.*, 130, 1650-1656.

Jorg, F., Schmitt, D. and Ziegahn, K-F. (1985). Materialschaden durch Luftverunreinigungen. *Fraunhofer-Institut fur Treib Und Explosivstoffe Report UBA-FB 84-106 08 010.*

Kaiser, W.D. (1977). *Korrosion*, 8, 3-15.

Keddie, A.W.C. and Williams, M.L. (1976). Estimates of contributions to smoke and sulphur dioxide concentrations during the high pollution episode in London, December 1975. *Paper to Standing Conference of Co-operating Bodies.*

Kenworthy, L. (1934). *BNFMRA Research Reports 231, 236, 254, 270, 293.*

Koeney, R.L. and Raiffa, H. (1976). *Decisions with multiple objectives.* New York: Wiley.

Koller, J. (1979). Causes of degradation of stone-like materials. *Bautenschutz u. Bausanierung*, 2, 8-12.

Kucera, V. (1976). Effects of sulphur dioxide and acid precipitation on metals and anti-rust painted steel. *Ambio*, 5, 243-248.

Kulis, M. (1979). Technisch-okonomische auswertung der korrosionsverluste an stahlkonstruktionen im Gebiet Nordbohmen. *Werkstoffe u. Korrosion*, 28, 197-205.

Larsen, R.I. (1969). A new mathematical model of air pollutant concentration averaging time and frequency. *JAPCA*, 19, 24-30.

Latimer, K.G. and Booth, F.F. (1963). Atmospheric exposure of aluminium alloys for three years in Nigeria. *Metallurgia*, 67, 73-80.

Layton, W.N. (1965). *Trans.Inst.Met.Fin.*, 43, 153.

Leary, E. (1983). *The Building Limestones of the British Isles.* London: HMSO.

Lewis, F., May, E. and Bravery, A.F. (1985). Isolation and enumeration of autotrophic and heterotrophic bacteria from decayed stone. In the proceedings of the Vth *International Congress on Deterioration and Conservation of Stone* (Lausanne). 633-642.

Liddiard, E.A.G. and Whittaker, J.A. (1961). Atmospheric corrosion behaviour of two structural aluminium alloys (H10 and H15). *J.Inst.Met.*, 89, 423-428.

Lifka, B.W. (1977). Paper to Western Metal and Tool Exposition and Conference, WESTEC, LA.

Liss, P.S. (1971). Exchange of SO_2 between the atmosphere and natural waters. *Nature*, 233, 327-329.

Livingston, R.A. and Baer, N.S. (1983). Mechanisms of air pollution-induced damage to stone. In the proceedings of the VIth *World Congress on Air Quality.* 33-39. Paris: IUAPPA.

Livingston, R.A (1985). The role of nitrogen oxides in the deterioration of carbonate stone. In the proceedings of the Vth *International Congress on Deterioration and Conservation of Stone* (Lausanne). 509-516.

Mackay, M.D., Beckman, R.J. and Gonover, W.J. (1979). *Technometrics*, 21, 239-245.

Mansfeld, F. (1980). Regional air pollution study: Effects of airborne sulphur pollutants on materials. *Environmental Protection Agency (USA) Report 600/4-80-007.*

Mansfeld, F. (1982a). New approaches to atmospheric corrosion resistance using electrochemical techniques. In *Corrosion Processes (Parkins, R.N., Ed). 1-76. London: Applied Science Publications.*

Mansfeld, F. (1982b). Electrochemical methods for atmospheric corrosion studies. In *Atmospheric Corrosion (Ailor, W.H., Ed).* 139. New York: John Wiley and Sons.

Martin, A. and Barber, F.R. (1987). Measurements of material lost from limestone specimens to rain. *Central Electricity Generating Board Report OED/STM/87/00003/N.*

Mattson, E. and Holm, R. (1968). Copper and copper alloys. *ASTM STP*, 435, 187-210.

Mazurkiewicz, B., Banas, J., Beida, K. and Piotrowski, A. (1976). *Zesz.Nauk.Akad.Gorn. – Hutn. im Stanislawa Staszica, Mat.Fiz.Chem.*, 27, 115.

McFadden, J.E. and Koontz, M.D. (1980). Sulphur dioxide and sulphate material study. Report prepared for the Environmental Protection Agency (USA) by Geomet. *Geomet Report* ES-812.

McGeary, F.L. *et al.* (1968). Atmospheric exposure of non-ferrous metals and alloys – Aluminium: seven-year data. *ASTM STP*, 435, 141-173.

McKenzie, M. (1982). The corrosion performance of weathering steel in highway bridges. *Atmospheric Corrosion* (Ailor, W.H. Ed), 717-741. New York: John Wiley and Sons.

Merry, C.J. and LaPontin, P.J. (1985). Regression models for predicting building materials distribution in four north eastern cities. Special report prepared for the Environmental Protection Agency (USA). *US Army Corps of Engineers, Cold Region Research Laboratory Special Report* 85-24.

Mikailovsky, Yu.N., Strekalov, P.V. and Agafonov, V.V. (1980). A model of metal corrosion in the atmosphere based on meteorological and aerochemical factors. *Zash.Metallov.*, 16, 396-413.

Mikailovsky, Yu.N. (1982). Theoretical and engineering principles of atmospheric corrosion of metals. In *Atmospheric Corrosion* (Ailor, W.H. Ed), 85-105. New York: John Wiley and Sons.

Ministry of Agriculture, Fisheries and Food (MAFF) (1986). *United Kingdom atmospheric corrosivity values.* (2nd edition).

Mishan, E.J. (1982). *Cost-Benefit Analysis.* Oxford: Oxford University Press.

Muller-Golchert, W. (1982). Acid attack from the air. *Mappe*, 210, 772-776.

Munier, G.B., Posta, L.A., Reagot, B.T., Russiello, B. and Sinclair, J.D. (1980). Contamination of electronic equipment after an extended urban exposure. *J.Electrochem.Soc.*, 127, 265-272.

National Trust (1983). *Properties Open to the Public.* London: National Trust.

Newton, R.G. (1982). *The deterioration and conservation of painted glass: a critical bibliography.* Oxford: Oxford University Press for the British Academy.

Newton, R.G. (1986). Does acid rain affect stained glass? In the *House of Commons Environment Committee First Special Report* (Follow-up to the Environment Committee Report on Acid Rain). 31-32.,

North Atlantic Treaty Organisation (NATO) (1985). *Atmospheric measurements of air pollution – Pilot study on conservation and restoration of monuments* (Aff/S 598-85, No.158). Germany: NATO.

Novakov, T., Chang, S.G. and Harker, A.B. (1974). Sulphates as pollution particulates: Catalytic formation on carbon (soot) particles. *Science*, 186, 259-261.

Nriagu, J.O. (1978). *Sulfur in the Environment.* New York: John Wiley and Sons.

Olimsted, J. (1982). The atmospheric corrosion of aluminium. *M.Sc thesis, University of Manchester Institute of Science and Technology.* (Unpublished).

Organisation for Economic Co-operation and Development (OECD) (1972). *The costs and benefits of sulphur oxide control.* Paris.

Organisation for Economic Co-operation and Development (OECD) (1981). *The costs and benefits of sulphur control. A methological study.* Paris.

Organisation for Economic Co-operation and Development (OECD) (1984). *International comparison study on reactor accident consequence modelling.*

Parker, W. (1881). *J.Iron and Steel Inst.*, 18,89.

Patzelt, H., Tunker, G. and Scholze, H. (1985). Untersuchungen zum schutz mittelalterlicher glasfenster ergansende versuche. *Umweltforschungsplan des Bundesministers des Innern, Forschungsbericht* 106 08 005/01. Berlin.

Payrissat, M. and Beilke, S. (1975). Laboratory measurements of the uptake of sulphur dioxide by different European sites. *Atmos.Environ.*, 9, 211-217.

Perrin, D.A. (1984). Dispersion modelling of concentrations of nitrogen oxides in London. *Warren Spring Laboratory Report* PRC 110.

Perrin, D.A. (1986). Modelling the transport and removal of sulphur dioxide emissions in the United Kingdom. *Warren Spring Laboratory Report* LR 560 (AP)M.

Photochemical Oxidants Review Group (PORG) (1987). *Ozone in the United Kingdom.* AERE Harwell.

Plummer, L.N., Parkhurst, D.L. and Wigley, T.M.L. (1979). In *Chemical Modelling in Aqueous Solutions*, ACS Symposium Series 93. Washington: American Chemical Society. 537-573.

Pitty, A.F. (1968). The scale and significance of solutional loss from the limestone tract of the Southern Pennines. *Proc.Geol.Assoc.*, 79, 153-177.

Price, C.A. (1981). Brethane stone preservation. *Building Research Establishment Report* CP1/81.

Price, C.A. and Ross, K.D. (1984). The cleaning and treatment of limestone by the 'lime method'. Part II. A technical appraisal of stone conservation techniques employed at Wells Cathedral. *Monumentum*, 301-312.

Programme Analysis Unit (1972). *An economic and technical appraisal of air pollution in the United Kingdom.* AERE Harwell.

Property Services Agency (PSA) (1982). *PSA Historic Buildings Register.* Volume 1 Northern and Eastern England. London: HMSO.

Property Services Agency (PSA) (1983). *PSA Historic Buildings Register.* Volume II London. London: HMSO.

Property Services Agency (PSA) (1984). *PSA Historic Buildings Register.* Volume III Southern England. London: HMSO.

Reddy, M.M., Sherwood, S. and Doe, B. (1985). Limestone and marble dissolution by acid rain. In the proceedings of the Vth *International Congress on Deterioration and Conservation of Stone* (Lausanne). 517-525.

Review Group on Acid Rain (RGAR) (1983). *Acid deposition in the United Kingdom.* Warren Spring Laboratory.

Review Group on Acid Rain (RGAR) (1987). *Acid deposition in the United Kingdom.* (2nd Report). Warren Spring Laboratory.

Richards, L.W. (1983). Comments on the oxidation of NO_2 to nitrate – day and night. *Atmos. Environ.*, 17, 397-402.

Rogers, F.S.M. (1984). A revised calculation of gaseous emissions from UK motor vehicles. *Warren Spring Laboratory Report* LR 508 (AP).

Rosenberg, H.S. and Grotta, H.M. (1980). *Env.Sci. and Tech.*, 14, 470-472.

Rosenfield, G. (1984a). *Cluster Sampling Design for Building Surveys.*

Rosenfield, G. (1984b). Spatial sampling for building materials. In The Changing Role of Computers in Public Agencies (Schmitt, R.R. and Smolin, H.J., Eds). *Proceedings of trhe Annual Conference of the Urban and Regional Information Systems Association August 12-15 1983, Seattle WA.*

Rozenfeld, I.L. (1972). *Atmospheric Corrosion of Metals.* NACE.

Salmon, R.L. (1970). Systems analysis of the effects of air pollution on materials. *U.S. DHEW, PHS, National Air Pollution Control Administration. Final Report* PH-22-68-35.

Saunders, K.G. (1981). Effects of air pollution on materials – an international debate. *Clean Air*, 11, 124.

Saunders, K.G. (1983). Air pollutant effects on materials. In *Proceedings of the VIth World Congress on Air Quality.* 59-66. Paris: IUAPPA.

Schaffer, R.J. (1932). The weathering of natural building stones. *DSIR Building Research Special Report* 18. (Reprinted 1972, 1985).

Schmid, E.V. (1984). Performance characteristics of organic coatings during weathering. *Applica*, 91 (13) 7-14; (14) 7-11.

Scholes, and Jacob (1970). *Int.Symp. Copper and Its Alloys, Amsterdam.*

Sharp, A.D., Trudgill, S.T., Cooke, R.U., Price, C.A., Crabtree, R.W., Pickles, A.M. and Smith, D.I. (1982). Weathering of the balustrade of St. Paul's Cathedral, London. *Earth Surface Processes and Land Forms*, 7, 387-389.

Shaw, T.R. (1978). Corrosion map of the British Isles. *ASTM STP*, 646, 204-215.

Shrier, L.L. (1976). *Corrosion.* Butterworth.

Simpson, D. (1986). An analysis of nitrogen dioxide episodes in London. *Warren Spring Laboratory Report* PRC 110.

Smith, K. (1975). *Principles of Applied Climatology.*

Skerry, B.S. (1980). Atmospheric corrosion of metals by smoke aerorols/hydrocarbons/SO_2 polluted systems. Ph.D. Thesis University of Manchester Institute of Science and Technology. (Unpublished).

Skerrey, E.W. (1982). Long-term atmospheric performance of aluminium and aluminium alloys. *Atmospheric Corrosion* (Ailor, W.H., Ed), 329-352. New York: John Wiley and Sons.

Spanton, A.M. (1983). Dispersion modelling of sulphur dioxide concentrations in London with reference to the EC Directive Limit Values. *Warren Spring Laboratory Report* LR 457(AP).

Spedding, D.J. (1969). Sulphur dioxide uptake by limestone. *Atmos.Environ.*, 3, 683-685.

Sperling, C.H.B. and Cooke, R.U. (1985). Laboratory simulation of rock weathering by salt crystallisation and hydration processes in hot, arid environments. *Earth Surface Processes and Landforms*, 10, 541-555.

Sugden, R. and Williams, A. (1978). *A Practical Cost Benefit Analysis.* Oxford: Oxford University Press.

Sweeting, M.M. (1966). The weathering of limestone. In *Essays in Geomorphology* (Drury, G. Ed). 177-210. Heinaman.

TRC (1983) Air pollution damage to man-made materials: Physical and economic estimates. *Electric Power Research Institute (EPRI) Report* EA-2837.

Thomas, T.M. (1970). The limestone pavements of the North Crop of the South Wales Coalfield, with special reference to solution and processes. *Trans.Inst.Brit.Geogr.*, 50, 87-105.

Truman, J.E. (1979). The atmospheric corrosion resistance of stainless steel. *Stainless Steel Ind.*, 35, 11-13, 17.

Umweltbundesamt (Hg.). (1980). *Luftverschmutzung durch Schwefeldioxid. Berlin: Bericht.*

United Nations Economic Commission for Europe (UNECE). (1982). Effects of sulphur compounds on materials, including historic and cultural monuments. *UNECE Report* ENV/1EB/R16.

United Nations Economic Commission for Europe (UNECE). (1984). *Air-borne sulphur pollution – Effects and control report prepared within the framework of the convention on long range transboundary air pollution.* New York: United Nations.

Van Egmond, N.D. and Kesseboom, H. (1985). A numerical mesoscale model for long-term average NO_x and NO_2 concentrations. *Atmos.Environ.*, 19, 587-595.

Vernon, W.H.J. (1923). First (experimental) report to the atmospheric corrosion research committee (of the British non-ferrous metals research association). *Trans.Farad.Soc.*, 19, 839-845.

Vernon, W.H.J. (1927). Second experimental report to the atmospheric corrosion research committee (British non-ferrous metals research association). *Trans.Farad.Soc.*, 23, 113-204.

Vernon, W.H.J. (1943). The corrosion of metals in air. *Chem. and Ind.*, 62, 314-318.

Vernon, W.H.J. (1957). Metallic corrosion and conservation. In *The Conservation of National Resources*. 105-133. London: Inst.Civil Eng.

Walton, J.R., Johnson, J.B. and Wood, G.C. (1982). Atmospheric corosion initiation by sulphur dioxide and particulate matter. *Br.Corros.J.*, 17, 59-64.

Williams, H. (1985). Identification and assessment of materials damage to buildings and historic monuments by air pollution. In *The Effects of Air Pollution on Historic Buildings and Monuments* (Commission of the European Communities). III-17 – III-68.

Williams, M.L., Bower, J.S., Maughan, R.A. and Roberts, G.H. (1981). The measurement and prediction of smoke and sulphur dioxide in the East Strathclyde region: Final report. *Warren Spring Laboratory Report* LR 412 (AP).

Willison, M.J., Clarke, A.G. and Zeki, E.M. (1985). Seasonal variation in aerosol concentration and composition at urban and rural sites in Northern England. *Atmos.Environ.*, 19, 1081-1089.

Winkler, E.M., (1970). Decay of stone. In *Conservation of Stone*, Vol.1. IIC New York. 1-14.

Winkler, E.M., (1973). *Stone: properties, durability in Man's environment.* Berlin and New York: Springer-Verlag.

Weatherley, M-L., Gooriah, B.D. and Charnock, J. (1975). Fuel consumption, smoke and sulphur dioxide emmisions and concentrations, and grit and dust deposition in the United Kingdom, up to 1973/4. *Warren Spring Laboratory Report* LR 214(AP).

Wise, A.F.E., Sexton, D.E. and Lillywhite, M.S.T. (1965). Air flow round buildings. *Proc. of Urban Planning Symp., Building Research Establishment, 8th January 1965*, 71-102.

Yocum, J.E. *et al* (1982). Review of air pollutant damage to materials. *Environmental Protection Agency (USA) Report* EPA-600/8-82-016.

Zakipur, S., Leygraf, C. and Portuoff, G. (1986). Studies on corrosion kinetics on electrical contact materials by means of a quartz crystal microbalance and XPS. *J.Electrochem.Soc.*, 133, 873-876.

ACKNOWLEDGEMENTS

Some of the experts who compiled this report also contributed to the Watt Committee Report on Air Pollution, Acid Rain and the Environment (Sub-Group 5) being prepared concurrently and this has led to some textural overlap between that section of the Watt Committee Report and this Report.

We should also like to acknowledge the help given by the members of the sub-groups in the preparation of this report.

Appendix A: Description of the Atmospheric Dispersion Models used in Chapter 5

A.1 SENSITIVITY, VALIDATION AND ACCURACY OF MEDIUM RANGE DISPERSION MODELLING

The processes that determine the transport and dispersion of releases into the atmospheric boundary layer are numerous and complex. Consequently, estimates of quantities, such as time averaged concentrations and deposition rates, are expected to be uncertain in the sense that they will differ from measured values. The conditions under which dispersion takes place are characterised by an inherent variability and randomness. This is evident, for example, in the diurnal and seasonal variations in meteorology, the variation in topography with location, and the changes in surface conditions such as state of vegetation cover.

This complexity has important implications for the choice of dispersion modelling methods. The relative advantages of two models, the widely adopted Gaussian plume type of model and the more involved eddy-diffusivity approach, are discussed in the report of the Atmospheric Dispersion Modelling Working Group (ADMWG 1979).

The Gaussian plume model is essentially a parameterisation of dispersion behaviour based on experimental results relating dispersion rates to readily measured meterological variables; it contains relatively little description of the physical processes involved.

The eddy-diffusivity approach offers a greater apparent flexibility in describing dispersion by turbulent motions, but a clear superiority in the reliability of the estimates produced has yet to be demonstrated. The models are usually computationally more time-consuming than the simpler Gaussian approach.

Differences between estimates obtained using the two methods can usually be reconciled as being within the range of uncertainty. For example, Barker (1978) carried out a detailed comparison between Gaussian and diffusivity based models, and concluded that for short averaging times (3 minutes) the predictions of the two models for ground level concentrations usually agreed to within 50% for ground level releases. For elevated releases and distances of the order of, or greater than, the distance to the maximum ground level, concentration agreement was to within a factor of three. The corresponding estimate of the accuracy of the models was within a factor of three. Barker urged that where greater accuracy was required the predictions would need to be validated against experimental data.

Agreement between different models does not, by itself, demonstrate validity since the predictions of two models might agree very closely with each other and yet still differ markedly from measured concentrations. For longer averaging times, say quarterly or annual, one may expect a closer correspondence than for the cases considered by Barker. Such a validation exercise was reported by Spanton (1983) in relation to winter mean SO_2 concentrations in London. For the 1979/80 winter 53% of the estimated concentrations were within $\pm$ 25% of the measured values, and 82% were within $\pm$ 50%. The situation considered here is more complicated in that multiple sources are involved, and Spanton considered that errors in the emission data were probably the most important source of discrepancies between the estimated and measured values.

A further aspect of modelling which needs to be considered is the sensitivity of the estimate to variations in values of user-selected variables or internal parameters of the model. A valid model will be sensitive to such factors in the same way as the actual mechanisms that are being described in the model. However, in the case of a complicated system, such as the atmosphere, it may not be practicable to provide a detailed and systematic validation of the various components of the model. An important technique that is useful in exploring the degree of consensus between modelling methods lies in the use of benchmark comparisons and sensitivity studies. In an international study using dispersion modelling techniques basically the same as those used in this exercise but applied to nuclear reactor accidents, the OECD (1984) reported a high degree of consensus in many respects between different models, but large discrepancies in some important aspects. Whilst the relative emphasis could be expected to be different in the case of acid emissions, similar international comparison in this area may well be merited.

The considerations described above suggest that in some cases modelling estimates may not be able to provide unequivocal attribution of observed ground level concentrations. This will be the case particularly where the apparent attribution suggests a fairly even distribution, rather than the dominance or insignificance of one of the source types. In such cases a scheme of attribution would probably have to depend on experimental validation. Whilst it is possible to envisage ways in which this might be done, for example, by doping one type of sulphur source with a distinguishing marker, such costly experiments would need to be fully justified.

A.2 THE CENTRAL ELECTRICITY RESEARCH LABORATORY (CERL) MODEL

A.2.1 The Short Range (Urban) Model

CERL have developed a fairly simple approach to modelling the dispersion of pollutants over urban scales. A more complex approach, where every individual source in an urban area is modelled separately, is not practicable given the lack of sufficiently well resolved emission data.

The CERL approach is to lump all the emitters together and assume that dispersion is governed by a limited number of metereological parameters. This approach has taken to its extreme by Gifford and Hanna (1973) in proposing a box-model type formula,

$$X = c \frac{q}{u} \qquad \text{Eq A.1}$$

where x = annual average pollutant concentrations
q = the emission rate of pollutant per unit area within the city
u = the mean surface wind speed
c = a coefficient ($\sim$50 for sulphur oxides).

A similar formulation has been used by Bennett et al (1986) to interpret survey results from Central London. In this case a power-law profile was assumed for the wind speed and a Gaussian profile for pollutant concentration at a given distance from the source. (It is assumed that the pollutant is emitted close enough to the ground to be entrained in the urban boundary layer.) Given a simple dependence of the vertical standard deviation of material in the plume, σ_z, on distance downwind, it is possible to integrate over all distances between the city centre and the periphery and to obtain a formula of the form:

$$X = A(R) \frac{q}{u_g} \qquad \text{Eq A.2}$$

Where R = radius of the area of mean emission density q, and u_g is the geostrophic windspeed. The formula is valid over periods of a few hours. The range for values of A(R) (Table A.1) show that for radii greater than 10 km, the coefficient increases only slowly as the city size increases; this was the rationale given by Gifford and Hanna for applying a single coefficient to all large cities. It should also be borne in mind that urban atmospheres tend to remain neutrally stratified whatever the conditions outside the city, and this leaves windspeed as the only important meteorological parameter in Equation A.2. The discrepancy between this coefficient and that proposed by Gifford and Hanna is consistent with the difference between the surface and geostrophic windspeeds. Temperature will only affect the emission term in Equation A.2. Analysis of survey results in Central London for the winter of 1983-84 (Bennett et al 1986) demonstrated the inverse dependence on wind speed, and a slight dependence on temperature.

Table A.1 Concentration coefficient, A(R), as a function of urban radius, assuming a power-law index, $\alpha = 1/7$ for wind speed, and $\sigma_z/x = 30$ mrad for the vertical plume speed (CERL Models).

R (km)	A(R)
1	31.3
2	48.3
3	59.7
5	75.1
10	96.4
20	117.3

Good SO_2 inventories exist for the London area (Ball & Radcliffe 1979; Ball & Armogie 1980) and these data may be used to predict annual average SO_2 concentrations in Central London. For 1978-79, Armogie and Ball (1983) divided London into 3 areas: a central statistical area (area 28.5 km², emissions density 15.5 μg SO_2 m^{-2} s^{-1}), the central Boroughs (area 143 km², emission density 1.8 μ_g SO_2 m^{-2} s^{-1}) and the rest of London (area 1,446 km², emission density 2.2 μg SO_2 m^{-2} s^{-1}). Inserting these values in Equation A.2 gives a mean value to the product $X\mu_g$ of 1,048 μg SO_2 m^{-2} s^{-1}.

The harmonic mean geostrophic windspeed for winter 1983-84 was calculated to be 9.4 m s^{-1} and dividing the product by this value gives a mean SO$_2$ concentration for 1978-79 in Central London of 111 μg m^{-3}. By comparison, the four National Survey sites in Central London (Lambeth 7, Lambeth 8, City 16 and City 17) returned mean values of 100 ± 23 μg SO$_2$ m^{-3}. Given that the observed values include contributions from external sources the simple calculation thus rather overestimates the urban contribution. Since emissions approximately halved between 1978/79 and 1983/84, the predicted urban concentration of SO$_2$ in the latter year would be 55 μg m^{-3}.

A.2.2 The Long Range Model

The model used in these calculations is that developed by Fisher and applied to calculate wet and dry deposition for the Acid Rain Review Group (RGAR 1983). The model sets out to calculate long-term averages by dividing the meteorological conditions into categories (5 of windspeed, 4 of stability); calculating dispersion functions for each of these categories; and then taking an ensemble mean of the contribution from each category.

The variation in concentration of SO$_2$ with distance depends on vertical dispersion, removal by wet and dry deposition, oxidation to sulphate, and the frequency with which the wind blows in a given direction. The occurrence of rain is treated as a random process which allows for different average durations of wet and dry periods. This introduces two parallel dispersion functions, g_d and G_w, where

$G_d(z,t)$ = Probability of pollutant emitted at t = 0 reaching a height z in a *dry* period after a travel time t.

$G_w(z,t)$ = Probably of pollutant emitted at t = 0 reaching a height of z in a *wet* period after a travel time t.

These two function are analytical solutions of the coupled differential equations

$$\begin{pmatrix} \dfrac{dG_d}{dt} \\[2ex] \dfrac{dG_w}{dt} \end{pmatrix} = \begin{pmatrix} K\dfrac{\partial^2}{\partial z^2} - k_d - \dfrac{1}{t_d} & \dfrac{1}{t_w} \\[2ex] \dfrac{1}{t_d} & K\dfrac{\partial^2}{\partial z^2} - k_w - \dfrac{1}{t_w} - \wedge_2 \end{pmatrix}\begin{pmatrix} G_d \\[2ex] G_w \end{pmatrix} \quad \text{Eq A.3}$$

where K is the eddy diffusivity given by Table A.2 and

k_d = rate of oxidation of SO$_2$ to sulphate in dry periods = 1% h^{-1}
k_w = rate of oxidation of SO$_2$ to sulphate in wet periods = 1% h^{-1}
t_d = mean duration of dry periods = 70 h
t_w = mean duration of wet periods = 7 h
$\wedge_2$ = rate of removal of SO$_2$ by rain = 36% h^{-1}.

In the steady state, $\partial/\partial t = 0$ and $d/dt = u\partial/\partial x$. Lateral dispersion does not enter since it is taken account of by long-term averaging over wind direction.

Initially, G_d and G_w satisfy the condition

$$\begin{pmatrix} G_d(z,0) \\[2ex] G_w(z,0) \end{pmatrix} = \partial(z-h)\begin{pmatrix} P_d \\[2ex] P_w \end{pmatrix} \quad \text{Eq A.4}$$

where $P_w = 1 - P_d = .07$ is the probability of rain at the time of emission. The height, h, of high-level sources is the mean effective height of power station emissions; all emissions other than from power stations are assumed to be at ground level. If the emission height is greater than the boundary layer depth, then the source is assumed to make no contribution to ground-level concentrations for travel times of less than 3 h. At greater distances, Equation A.3 applies.

Table A.2 Eddy diffusivity and boundary layer depth according to stability category and windspeed (CERL Models).

Category Windspeed (ms^{-1})	Eddy Diffusivity (m^2 s^{-1})				Boundary Layer Depth (m)			
	1	2	3	4	1	2	3	4
2	52.50	.48	.12	.12	1500	120	60	200
6	52.50	4.32	1.08	1.00	1500	360	180	200
10	52.50	12.00	3.00	1.00	1500	600	300	200
14	52.50	23.52	5.88	1.00	1500	840	420	200
18	52.50	38.88	9.72	1.00	1500	1080	540	200

Boundary conditions to Equation A.3 are

$$K\frac{\partial G_d}{\partial z} = 0 \quad \text{at } z = a, \text{ the boundary layer depth,}$$

and

$$K\frac{\partial G_d}{\partial z} = v\,G \text{ at } z = 0, \qquad \text{Eq A.5}$$

where the dry deposition velocity, $v_d = 5$ mm s^{-1}. G_w obeys identical boundary conditions.

Having solved Equations A.3—A.5 for G_d and G_w, the long-term mean SO_2 concentration at a point is found by summing over all sources, i and all categories, j

$$\chi = \sum_i \frac{Q_i f_i}{2\,x_i} \sum_{u,\,j} \frac{P(u,\,j)}{u} \left(G_d(O, X_i/u) + G_w(0, x_i/u)\right)$$

$$\text{Eq A.6}$$

Q_i is here the emission rate of the ith source, x_i its distance from the receptor and f_i the relative probability of the wind blowing between the source and the receptor (Table A.3). $P(u, j)$ is the joint probability of a windspeed u together with stability category j (Table A.4). In the near field, these probabilities depend on the wind direction. They are based on measurements at the Belmont transmitter mast in the Lincolnshire Wolds.

Table A.3 Probabilities of different wind directions (CERL Models).

Direction	N	NE	E	SE	S	SW	W	NW
Weighting	.08	.10	.10	.12	.21	.21	.11	.06

Table A.4 Probabilities of windspeed and stability categories (CERL Models).

Category Windspeed (ms^{-1})	1	2	3	4
2	.020	.026	.032	.040
6	.059	.072	.073	.094
10	.046	.063	.056	.083
14	.022	.052	.045	.048
18	.008	.060	.053	.028

The calculations are made on a 20 km square grid and this inevitably leads to some loss of precision close to large sources. Emissions in the same square as the receptor are treated by placing them equally at the four corners of the square. In urban areas, the contribution from local sources will thus be substantially underestimated.

The model has been validated by a comparison of calculations and measurements of wet and dry deposition of sulphur over continental Europe

(Fisher 1983) and over the United Kingdom (RGAR 1983), and generally behaves well.

A.3 THE WARREN SPRING LABORATORY (WSL) MODEL

The model used by WSL in calculating the contributions to ground level concentrations (g.l.c.) of SO_2 at the sites of interest is in two parts—a short range (< 50 km) model applied to urban areas, and a longer range model which is used to calculate source-receptor contributions over distances greater than 50 km.

A.3.1 The Short Range Model

For dispersion over distances less than 50 km a long-period average Gaussian model is used. The model has been described elsewhere (Spanton 1983) and a detailed description will not be repeated here.

The model assumes that over long periods (eg annual averages) the wind direction fluctuations within each wind sector are uniformly distributed and hence that the pollutant concentrations are uniformly distributed within each sector so that the ground level concentration C at a downwind distance x from a point source is given by:

$$C(x) = \sqrt{\frac{2Q}{2\pi U\,\sigma_z\,2\pi x/n}} \;\exp\;\frac{-H^2}{2\sigma_z^2} \quad \text{Eq A.7}$$

where Q is the emission rate, U the wind speed, σ_z is the vertical standard deviation of material in the plume, and H is the effective stack height. The number of wind sectors n is taken in the WSL model to be 12.

Reflection of pollutant material by the ground or an elevated inversion is accounted for the by the virtual source method with the truncated series given below being used for an inversion height L

$$C(x) = \frac{1.5238Q}{U\sigma_z x}\left\{\exp\frac{-H^2}{2\sigma_z^2} + \exp\frac{-(H+2L)^2}{2\sigma_z^2} + \right.$$

$$\left. \exp\frac{(H-2L)}{2\sigma_z^2}\right\} \qquad \text{Eq A.8}$$

The effect of an inversion is included when the concentration at the inversion height is 10% of the plume centre-line concentration ie if $\sigma_z < 0.465L$ then Equation A.7 is used but if $\sigma_z > 0.465L$ then the concentration is given by Equation A.8.

For $\sigma_z > L$ the pollutant is assumed to be uniformly mixed in the boundary layer, and the equation used is

$$C(x) = \frac{1.9099Q}{UL\,x} \qquad \text{Eq A.9}$$

For area sources, the total emission within each grid square has been treated as a virtual point source located upwind of the centre of the square. The displacement distance is a weighted value depending on the cross-wind dimension of the grid square and is 2.321d where d is the length of side of the square. The equations used for area sources are thus A.7, A.8 and A.9 with x replaced by (x + 2.321d). An area source in this model can contribute to the concentration at a receptor on more than one wind direction, the criteria for contribution being given in detail in Keddie *et al* 1978.

When the receptor lies within the square whose emissions are being modelled, these emissions are assumed to be evenly distributed over the square and the contribution from the square is calculated by a box model

$$C = \frac{Q}{1.244\,dU\,H_b} \qquad \text{Eq A.10}$$

where 1.244d is a value, weighted for wind direction, dependent on the square side, d, and H_b is the height to which pollutants are mixed in the box. In this formulation an area source contributes to a receptor within it equally on all wind directions.

The equations above are applied to a matrix of six stability classes (A to F/G) for each of the 12 wind sectors. In major urban areas (eg London), in order to allow for the effect of the urban heat island, stability categories which are calculated on the basis of airport data (generally rural) to be Pasquill E, F or G, are each shifted one category towards neutral; so that the E categories become D, F becomes E etc. For each stability/wind sector combination, a harmonic mean wind speed is calculated and used in the equations A.7—A.10. The resulting concentration is then weighted by the appropriate frequency of occurrence of the stability/wind direction (sector) combination.

The parameterisation of σ_z used for all non-power station sources is that of Pasquill-Gifford with a modification due to Smith (1973) to allow for the effect of surface roughness. The initial spread of the plume due to building entrainment, downdash etc, is taken into account by means of a constant addition to $\sigma_z(x)$. For power stations the σ_z values from the Brookhaven experiments for a 100 m high release are used (Singer & Smith, 1966).

For point sources other than power stations, Briggs' plume rise equations are used (Briggs 1969) while Moore's plume rise formulae are used for power stations (Moore 1974).

A.3.2 The Long Range Model

If a receptor is 50 km from a source, the concentration of sulphur at the receptor is calculated using a model developed by Scriven and Fisher (1975a; 1975b) and Fisher (1975; 1978). The reader is referred to these papers for a detailed description, only the main features are described here.

In the vertical, dispersion is governed by an eddy diffusivity. In Fisher's original formulation (1975) the diffusivity was assumed to vary with mixing height, and windspeed. In this work the mixing height is taken as constant at 800 m as recommended in Fisher (1978) so that the diffusivity is a function of windspeed alone. On emission a constant fraction of SO_2 is assumed to be converted to sulphate (SO_4), thereafter a constant conversion rate is used. The removal of SO_2 by dry deposition is represented by a deposition velocity whilst dry deposition of SO_4^{2-} is neglected. Removal by precipitation is modelled using the technique of Rodhe and Grandell (1972) whereby the start of a wet (or dry) period is assigned a constant transition probability. The values used for the mean lengths of wet and dry periods and the washout coefficient are those used by Smith (1981). Variation in rainfall across the United Kingdom is represented by defining different rainfall regions as described by Fisher (1978).

Meteorological conditions are represented by a geostrophic windrose of twelve 30° sectors, with four categories of windspeed (Fisher 1978). Emissions and transport are taken always to occur in neutral stability. It is assumed that the plume of material from each source is uniformly distributed in each horizontal sector. At each receptor the concentration of SO_2, sulphate aerosol, and deposition of sulphur (wet and dry) is calculated by summing over all sources weighted for the different wind speeds and directions.

Values of the parameters used are shown in Table A.5

Table A.5 Parameters used in long range model (WSL).

Windspeed Range (m s^{-1})	0–5	5–10	10–15	15
Eddy diffusivity (m^2 s^{-1})	4	12	20	32
Mixing height (m)			800	
Fraction of SO_2 emitted as SO_4^{2-}			0.05	
Oxidation rate (s^{-1})			10^{-6}	
Deposition velocity for SO_2 (mm s^{-1})			8	
Mean duration of dry period (h)			8	
Mean duration of wet period (h)			40	
Washout coefficient (mm^{-1})			0.34	

Table A.6 Summary of results of sensitivity analysis (WSL Models).

Test Parameter	Description	Percentage Change in Calculated Concentrations (Averaged over 23 Sites)

0. Base Case	Pasquill-Gifford σ_z
	Briggs' Plume Rise
	Urban Adjustment: E, F/G Categories
	23 Sites:
	  Ave. Observed = 190 μg m^{-3}
	  Ave. Calculated = 194 μg m^{-3}
		0

1. Urban Adjustment	Non	+4
2. ,, ,,	All Categories Modified	−24
3. Stability Categories	All Neutral	+6
4. σ_z Formula	Smith's σ_z Urban Adjustment: E, F/G Categories	−10
5. ,, ,,	Smith's σ_z No Urban Adjustment	+20
6. ,, ,,	Smith's σ_z Urban Adjustment: All Categories	−37
7. σ_z & Plume Rise Formula	Brookhaven σ_z } For Point Sources Moore's Plume Rise }	Negligible
8. σ_z^0	Area Sources, Categories E, F/G reduced to 5 m	+1
9. H_e	Comm/Ind Area in Square A - doubled	−10
10. H_b	All reduced by 25%	+8
11. H_b	Categories E, F/G reduced by 50%	+4
12. z_9	Reduced from 1.0 m to 0.3 m	+6
13. L	Categories A, B, C, D increased to 2000 m	−1

A.3.3 Sensitivity and Accuracy of the Models

Considering first the short range model, a sensitivity analysis was carried out for the application of the model to calculations in the Greater London area (Spanton 1983) and the results are shown in Table A.6. The most sensitive dispersion parameter is the vertical dispersion coefficient σ_z, particularly when this was changed in conjunction with changes in the urban heat island adjustment described above. It is also interesting to note that for the London calculation, assuming all stability categories to be neutral only changed the overall average concentration by +6%.

WSL studies of Gaussian dispersion models in practical applications to urban (short range) areas suggest that given an acceptably good emission inventory, pollutant concentrations could be calculated to within ± 30% with a confidence of 60—90%.

An evaluation of the long range model has been carried out by Perrin (1986) who used the model to calculate gaseous SO$_2$ concentrations and wet-deposited sulphate over the whole of the United Kingdom. In terms of the prediction of gaseous SO$_2$ concentrations, an analysis of the frequency distribution of errors analogous to that already described for the urban model, gives 65% of remote SO$_2$ sites (N = 14) with errors between observed and calculated of ± 30%, with 80% of sites ± 50%.

A.4 THE BRITISH COAL CORPORATION (BCC) MODEL

A.4.1 Model description

Standard dispersion modelling methods have been utilised, based largely on techniques recommended in reports by the Atmospheric Dispersion Modelling Working Group (1979; 1981a; 1981b; 1981c; 1983), the GLC Scientific Services (Ball 1980) and Warren Spring Laboratories (Spanton 1983).

Two basic models were used which deal with large point sources (ie power stations), and with regional or local area sources. Both models use the Gaussian dispersion equation for continuous releases of pollutant. This allows the dispersion of pollutant downwind from the source within the atmospheric boundary layer (which is typically 1 km thick). The dispersion formula allows 'reflections' from both the ground and the inversion lid of the boundary layer. Both the ground and the lid are assumed to be perfect reflectors. The formula for the spread of pollutant in the horizontal and vertical crosswind directions use Pasquill's dispersion coefficients (ADMWG 1979; Ball 1980). Estimates of dry deposition of SO_2 en-route are incorporated in the models using the source depletion method (ADMWG 1981a; Rote 1980). This decreases the source strength at 5 km intervals using a simplified Gaussian dispersion model and a deposition velocity of 0.8 cms^{-1} (Garland 1978).

In the point source model, Moore's (1974) equations are used to make allowance for plume rise above the stack caused by the buoyancy and momentum of the flue gases. For the area source model, runs at various fixed effective heights have been used to estimate the contributions from domestic, commercial, and industrial fuel use.

The calculations are undertaken for all combinations of four atmospheric stability classes and six windspeed classes at half degree intervals of wind direction. The results are weighted according to a frequency matrix of meteorological data. The weather matrix used is derived from 1983/84 data RAF Finningley (40 km North-West of Lincoln) and they provide frequencies for combinations of 16 wind directions, 4 stability classes and 6 windspeed classes. Atmospheric 'lid' heights are assumed to be constant within each of the stability classes (ADMWG 1979). The results, when aggregated, represent estimates of the contribution to the annual average SO_2 concentration at a given site from the various sources.

A.4.2 Input Data

Emissions of SO_2 from power stations for 1983/4, 51 CEGB coal-fired, 10 CEGB oil-fired and 3 SSEB, were estimated from data in the CEGB statistical year book, SSEB annual report and BCC sources. The retention of sulphur in power station ash was taken as 4%.

Various methods have been used to estimate and disaggregate emissions local to the target area and elsewhere in the United Kingdom in order to evaluate the likely range of contributions. Sources used include government statistics on United Kingdom SO_2 emissions (Digest of Environmental

Statistics), and estimated emissions inventory for 1977 in a CEGB report, and 1981 census data. Emissions in the vicinity of the target location (eg in Lincoln itself) have been represented by points on a 1 km grid. Other emissions are taken as point sources, representing county sized areas.

A.4.3 Sensitivity To Model Type and Input Parameters

The development and testing of the dispersion models has allowed assessment of the sensitivity of predictions to the details of the models and their input parameters.

(a) Both Models

Model Type. The results are most sensitive to the type of Gaussian model used. The change from a simple ground reflection model used in earlier work to a model incorporating multiple ground and lid reflections resulted in increases of about 70% in predicted concentration. In contrast, the simplifications obtained by limiting the number of reflection terms to 3 and assuming uniform vertical distribution beyond a critical distance, have an insignificant (< 1%) effect on the results.

The values given to the dispersion coefficients are also critical to the predicted values. The change to Pasquill's widely accepted parameters from those used initially caused decreases in estimates of 25—35%.

Dry deposition. Dry deposition of SO_2 en-route, when incorporated in the model, reduced the results by about 1% for nearby sources and 7% overall for power stations. For simplicity a ground reflection only Gaussian model was used in this calculation, which probably under-estimated the dry deposition losses by around a third.

Weather data. Sensitivity to the weather data has been analysed for Lincoln and Bottesford (power stations). In both cases the combinations of 1983/84 data from RAF Finningley and typical United Kingdom data given in the ADMWG report (1979) resulted in variation from −8% to + 35% of the chosen values. The ADMWG data appear more typical of average United Kingdom conditions, whilst the Finningley data have greater contributions from northerly and southerly directions.

(b) Point Source Model

Plume rise. Various methods of plume rise estimation (Ball 1980; Moore 1974; Briggs & Moore 1975; Moore & Lee 1982) were evaluated for the earlier simple ground reflection model. These resulted in a variation of −20% to +35% of results using the chosen method (Moore 1974) for nearby sources and negligible variation for very distant sources.

'Lid' penetration. Sensitivity to the simplifying assumption of the impenetrability of the inversion lid to the atmospheric boundary layer was tested by examining the predicted effects of temperature differences at the inversion of 0°C (ie no inversion), 1°C and 6°C on contributions from nearby power stations. The 1°C and 6°C inversions reduced the result for an impenetrable lid by less than 1%, whereas the unreasonably hypothetical 0°C lid reduced predicted concentrations by 22%. The use of average values for the lid (for each atmospheric stability category) meant that the plume was always emitted from the stack below the lid height. (The tallest stack in the country is 250 m at Drax whereas the lowest lid height, for stable conditions, was 300 m). In practice there may be occasions when the lid falls below the height of tall stacks, but these are considered relatively infrequent.

(c) Area Source Model

Inventories. Estimates of regional and local emissions are a critical part of modelling contributions of SO_2 concentrations. The various estimates used resulted in ranges of ± 12% for local sources and ± 10% for distant sources. However, the inaccuracies are probably much greater as all the inventories are based largely on population data not regional fuel use. Heterogeneity of source distribution is an especially serious source of error at the local scale. For example, one source at a rural site in Lincolnshire had emissions in 1983/84 which were 10% of the estimated total for the whole county.

Grid disaggregation. Local sources are assumed to be evenly spread over the built-up area and represented by points on a 1 km grid. The precise positioning of the grid and further disaggregation of the nearest sources produces variation of more than 10% in predicted local contributions.

Emission heights. The mixture of area source emission heights, taking account of plume rise, largely depends upon the relative importance of domestic (about 20 m), commercial (30—50 m) and industrial (50—80 m) sources. There is more than an order of magnitude separating predictions of local contributions assuming all 20 m or all 80 m sources, but reasonable mixes of emission heights give a range of ± 10%.

For distant sources the effect of emission height on predicted contributions is insignificant as the pollutant becomes fairly uniformly mixed in a vertical direction within the atmospheric boundary layer whatever the initial emission height.

A.4.4 Assumptions

A number of simplifications and approximations were necessary in the model and these are discussed.

Continuity of weather conditions. Weather conditions are assumed to be constant between sources and receptors. This is probably a good approximation for nearby sources, but is far from reality for distant sources. Any improvement in the model would require major re-structuring to introduce probabilities of changes in wind direction and strength, and changes in stability conditions. One major effect would be to reduce the artificial sensitivity to details of the weather data matrix.

Diurnal and seasonal variations. The model assumes constant emissions throughout the year, whereas in reality there is significant seasonal and diurnal variation in fuel use. The correlation between these variations and weather conditions means that small errors are introduced.

Wet deposition and other losses. No allowance is made for wet deposition or other losses due to transformation of SO_2 in the atmosphere. Consequently modelled concentrations are slightly higher than they should be.

B: Relative Contributions of Point Sources and Local Sources to Peak Urban Pollutant Concentrations in York

Modelling described in the body of the report has shown that local sources may contribute 13-20 μg m^{-3} to the arithmetic mean SO_2 concentration in York. Within a lognormal distribution, this is related to the geometric mean by

$$\tilde{\chi} = u \exp (\tfrac{1}{2} \sigma^2) \qquad \text{Eq B.1}$$

where $\tilde{\chi}$ is the arithmetic mean, u the geometric mean and σ the logarithmic standard deviation. Typical values of e^σ for hourly averages of urban pollution from WSL work are 2.3-2.8. A frequency of 1 h per annum is 3.81 standard deviations removed from the mean and it follows that the peak hourly concentration is a factor of $e^{3.81\sigma} = 20$-50 greater than the geometric mean. Applying Equation B.1, the model then predicts a peak hourly concentration from local sources in York of ~250-600 μg SO_2 m^{-3}.

Larsen (1969) suggests that the median, ie the geometric mean, pollutant concentration should be a simple power law function of the averaging time. In this case, standard deviations at different averaging times may be related by

$$\sigma' = \sqrt{\log (T/t') / \log (T/t)} . \sigma \qquad \text{Eq B.2}$$

where T is the total time for which measurements have been made. Applying this equation and assuming T = 1 year, we find that for daily averages $e^\sigma = 2.0$-2.4. These values are confirmed by measured data on SO_2 in the Yorkshire region in the United Kingdom. A frequency of a 1 day per annum is 2.94 standard deviations from the mean and it follows that the model predicts a peak daily concentration from local sources in York of 100-170 μg SO_2 m^{-3}.

The peak hourly concentration from power stations at York is determined by the largest nearby power station. For hourly averages, there should be no problems with plumes lining up since the three nearest power stations, Eggborough, Ferrybridge and Drax have a fairly wide-angular separation. Of the three, Drax should have the largest impact since it now has a capacity of 4000 MW, twice that of each of the other stations. According to the exponential distribution the frequency with which a given concentration is exceeded is given by

$$f (\chi) = f_o e^{-\chi/\chi_0} \qquad \text{Eq B.3}$$

where f_o is the frequency with which non-zero concentrations are observed and χ_0 is the arithmetic mean of non-zero concentrations. Both f_o and χ_0 are functions of averaging time but the arithmetic mean,

$$\tilde{\chi} f_o \chi_0 \qquad \text{Eq B.4}$$

is not. Two short-term surveys of SO_2 have been carried out by CERL in Central York and during the second of these, in March 1986, an episode occurred during which the plume from Drax was observed very frequently over several days. Interpreting the observed hourly concentrations in terms of Equation B.3, we find $f_o = 19\%$ and $\chi_0 = 114$ μg m^{-3}. Clearly, the observed frequency, f_o is peculiar to the period of measurements and an educated guess for its long-term value (taking account of the relative frequency of southerly winds, the frequency with which the plume is dispersed to the ground, and the angular width of a plume over an hourly averaging time) would be $f_o = 3.5\%$. If, however, the observations are assumed to be a fair sample of what happens when the Drax plume is observed at York, it follows that we have a good estimate of χ_0 and the long-term mean SO_2 concentration at York from Drax is therefore

$$f_o \chi_0 = 4 \ \mu\text{g m}^{-3}$$

According to Equation B.3, the maximum hourly average in a year is given by

$$\chi_m = \chi_0 \log (8766 \ f_o) \qquad \text{Eq B.5}$$

and using the above values this amounts to 650 μg SO_2 m^{-3}. By comparison, the highest hourly concentration observed during the episode in March 1986 was 450 μg SO_2 m^{-3}. Thus, although the predicted annual mean from the power station is less than a third of that from local sources, the predicted maximum hourly concentration could be up to a factor of three larger.

Simpson *et al* (1984) have analysed SO_2 measurements around an isolated point source and suggest that the maximum annual concentration should be an inverse power law function of averaging time. From their data, it appears that the index of the power law should be about .62. Thus the maximum daily average should be a factor of $24^{.62} = 7.2$ smaller than the maximum hourly average, ie 90 μg m^{-3} for Drax at York. This is comparable with or less than the highest predicted daily mean from local sources.

For daily means, of course, Equation B.3 breaks down since many sources may contribute as the wind changes direction. The maximum daily average at York from all United Kingdom power stations will thus be rather higher than that predicted from Drax alone.

C: Guidelines for Setting Up and Maintaining National Materials Exposure Programme (NMEP) Sites

C.1 INTRODUCTION

This note presents the guidelines agreed upon by the BERG II sub-committee for the setting up and monitoring of sites within the National Materials Exposure Programme (NMEP). For useful results to be obtained from a nationwide monitoring network it is essential that standardised procedures are followed by all participating organisations.

C.1.1 The Network

The NMEP network aims to investigate materials damage at 29 sites across Britain. Sites in the network will contain atmospheric pollution monitoring equipment and a range of materials samples. The monitoring project is planned to last at least four years. Pollution monitoring equipment will require weekly site visits. Materials samples will require visits to remove some samples after one, two and four years of the project, with other samples remaining after this time for planned longer-term studies.

Two main types of sites are operating within the NMEP network, ie sites on open ground and sites on buildings. It includes a range of buildings of both traditional and modern construction. Sites differ in the degree of sophistication of monitoring equipment and materials samples present. This hierarchy is more fully explained in Section C.4.

C.1.2 Summary of Site Procedure

(i) Site selection, to conform as far as possible with guidelines for site requirements in Section C.2. Some sites were already established.

(ii) Installation of meteorological and pollution monitoring equipment, to conform as far as possible with requirements in C.4.

(iii) Construction of racks for materials samples, preparation and deployment of samples (to conform as far as possible with requirements in Sections C.5.1, 5.3, 5.4 and 5.5).

(iv) Initial site documentation. Record all relevant details of site (see Section C.3) and send a copy to the Building Research Establishment.

(v) Weekly measurement using meteorological and pollution equipment (see Section C.4.3).

(vi) Calibration of meteorological and pollution monitoring instruments as often as appropriate (in accordance with the manufacturer's recommendations).

(vii) Removal of materials samples for analysis, and deployment of replacement samples at the end of 1, 2 and 4 years (See Section C.5.6).

C.2 SITE REQUIREMENTS

Sites on buildings cannot be expected to meet the same criteria as open sites and two sets of site requirement guidelines have been prepared. All sites within the network must (as far as possible) fulfil the basic criteria for either open sites or sites on buildings.

C.2.1 Open Sites

* Local sources of contamination must be avoided (eg heating systems of buildings, road traffic). As a general guideline, there should be a horizontal separation of at least 100 m between the exposure site and any small point or mobile source.

* Wind shadowing of precipitation collectors and materials exposure racks should be avoided.

Hence:—

* No object should be closer to the precipitation collector or materials exposure racks than 1.5 times the width or 3 times the height the object extends above them.

* The precipitation collector should be installed over undisturbed land with the collecting funnel 1 to 2 m above the ground cover.

* Land use changes should be avoided. Ideal sites are those where no changes in land use, construction of new sources or obstruction, or substantial changes in vegetation are expected over the next four years.

* The influence of local topography should be minimised, as valley or ridge sites will be influenced by strong gusts of wind, which will affect the amount of precipitation collected.

* Sites should be accessible in all weather conditions to ensure a data collection rate of > 90%.

C.2.2 Sites on Buildings

* Sites should ideally be chosen on buildings which are likely to remain under the same ownership over the duration of the programme, and which are not going to be demolished or altered.

* Sites should be as far as possible from local sources of contamination (eg heating system flues).

* There should be a horizontal surface available which is large enough to contain all the monitoring equipment, easily accessible, and relatively secure against vandalism.

* Ideally, sites should be located at a height of approximately 10 m above ground level.

* The site should be as near as possible to the pollution monitoring equipment. Ideally the maximum distance between the exposure site and the pollution monitoring equipment should not exceed 5 m; ideally the monitoring equipment should be all on the same site.

* The site should be as safe and secure as possible (ie it should be easy and safe to work on the site). All equipment left on the site should also be securely fixed in position. All other safety standards should be adhered to (electrical, lightning conductors, etc).

C.3 SITE DOCUMENTATION

At each site within the NMEP network a record must be kept of various site characteristics. This record must be updated should any changes occur in the details during the monitoring programme. The following details should be recorded:—

* Site location on an Ordnance Survey map (full 8 figure grid reference).

* Photographs of the site and of the surroundings (a minimum of 1 view of the site + 1 view of the equipment on site).

* A complete listing of all materials exposed (see Section C.5).

* A complete listing of all meteorological and pollution monitoring equipment on site (see Section C.4).

* A full description of the site with reference to the guidelines set out in Sections C.2.1 and 2.2 above.

* A record of any other atmospheric pollution monitoring stations nearby, including details on their location, ownership and monitoring techniques.

* A record of any other materials degradation monitoring taking place at the site.

* Location of the nearest meteorological station which supplies at least wind speed and rainfall data.

A copy of these details should be sent to the Building Research Establishment. Any changes in the above details during the monitoring programme should be recorded in a similar manner (with dates).

C.4 METEOROLOGICAL AND POLLUTION MEASUREMENT

The following list indicates the minimum requirements for meteorological and pollution measurements at each site. The NMEP network will contain two levels of sites, defined according to the type of meteorological and pollution information collected, ie:—

Level 1 sites: Minimum level of meteorological and pollution measurements and at least the minimum range of materials exposed (see Section C.5).

Level 2 sites: Minimum level of meteorological and pollution measurements PLUS some or all of the DESIRABLE measurements and at least the minimum range of materials (see Sections C.4.2 and 5.2).

C.4.1 Minimum Requirement – Level 1 Sites

Variable	Method of Measurement	Frequency of Measurement
Relative humidity	Any acceptable method	Continuous (15 mins averages)
Wind speed and direction	Any acceptable method	Continuous (15 mins averages)
Air temperature	Any acceptable method	Continuous (15 mins averages)
SO_2	SO_2 bubbler or continuous monitor	Daily or 15 mins averages
Rainfall pH	WSL design bulk precipitation collectors + pH measurements suitable for rainwater	Weekly
NO_2	Diffusion tubes or continuous monitor (NO_x)	Weekly in urban areas/fortnightly elsewhere or 15 mins averages

C.4.2 Desirable Measurements – Level 2 Sites

Variable	Method of Measurement	Frequency of Measurement
Insolation/u.v.	Any acceptable method	Continuous
Rainfall intensity	Tipping bucket rain gauge	Every 0.2 mm of rainfall
Wind-driven rainfall	BRE wall gauge	Weekly
Wind-driven rainfall intensity	Tipping bucket rain gauge	Every 0.2 mm of rainfall
Ozone	Continuous monitor	Continuous (15 mins average)
Smoke	Whatman No. 1 filter before SO_2 bubbler	Daily
Rainfall composition	WSL design bulk precipitation collectors, analysis of all major ions and conductivity, or wet-only collector and similar analyses	Weekly
		Daily
Time of wetness	Any acceptable method	Continuous
NO_x	Continuous monitoring	Continuous (15 mins average)

C.4.3 Measurement Procedures

The following procedural guidelines should be followed at all sites:—

* Weekly measurements should preferably be taken on Tuesdays, Wednesdays or Thursdays.

* Results should be recorded and a copy sent to the Building Research Establishment.

* Regular comparisons between the measurements taken at each site should be made to provide quality control. Intercomparisons will be made every 6 months, starting soon after the programme begins.

* Continuous monitoring data should be recorded every 15 minutes.

* Changeover on SO_2 bubblers should be 12 noon.

* All times should be recorded as GMT.

C.5 MATERIALS EXPOSED AND MONITORING METHODS

C.5.1 Minimum Requirements – Level 1 Sites

The following materials have been agreed upon to form a minimum requirement for each site:—

Material†	Sample Type	No of Samples‡
Mild steel (BS 1449, selected by BSC to be comparable in minor elements with steel for ISO trial)	Plate*	12
Aluminium (BS 1470)	Plate	12
Galvanised mild steel (BS 1449, hot dip galvanised to BS 729, average Zn weight 320 g m⁻²)	Plate	12
Alkyd paint system (Defence spec. TS 10266)	Plate	12
Portland limestone	Tablet **	24
White Mansfield dolomitic sandstone	Tablet	24

NB * Plates will be 75 mm × 50 mm × 3 mm
   ** Tablets will be 50 mm × 50 mm × 8 mm
   † Metal and paint samples will be exposed at an angle of 45 degrees and south facing wherever possible. Stone samples will be exposed on carousels.
   ‡ The sample numbers specified allow triplicate samples to be removed after 1, 2 and 4 years, with one group left for longer term monitoring. There are twice as many stone samples, these are deployed in both exposed and sheltered situations.

C.5.2 Desirable Materials – Levels 1 and 2 Sites

Material	Sample Type	No of Samples
Copper (BS 2870)	Plate	12
Monk's Park limestone	Tablet	24

The Department of the Environment will be responsible for provision and analysis of all samples. A reference set of all samples will be kept at the Building Research Establishment.

C.5.3 Sample Preparation

The following guidelines should be followed when preparing samples for the monitoring sites:—

(a) *Metal and paint system samples*

* Metal samples must be cleaned and prepared as follows:—

  (i) Mild steel. Condensing solvent (trichlorethane) cleaned. Blast cleaned to Swedish spec. 2.5 (two and a half).

  (ii) Aluminium, galvanised steel, alkyd paint system, and copper. Condensing solvent (trichlorethane) cleaned.

* Plastic gloves should be worn at all times when weighing and handling samples.

* Metal samples should be weighed to the 5th significant figure and these weights recorded.

* Metal samples should be numbered with a plastic tag tied to the sample. The tie should be loosened during exposure to minimise crevice formation.

(b) *Stone samples*

* Plastic gloves should be worn at all times when handling stone samples.

* Stone samples must be cleaned and prepared as follows:—

  (i) Samples brushed and rinsed with distilled water, then air dried.

  (ii) Samples dried at 105 degrees C for 24 hours, cooled in a desiccator, and brushed again before weighing.

* Stone samples must be weighed to an accuracy of 0.05 g and the weights recorded.

* Stone samples should be identified using a numbered PVC label, attached with stainless steel wire to the arm of the carousel upon which they are mounted (see Section C.5.5).

(c) *All samples*

* All samples should be photographed on emplacement at the site and after each exposure period. Photographs should be taken with a 35 mm camera, equipped with a 50 mm lens, and using colour film.

* A detailed plan of the disposition of all samples on each site should also be drawn up.

C.5.4 Sample Numbering

The following numbering scheme has been agreed upon:—

* Each sample will be identified using a number to identify the site, a reference code for the material type, and a sample number (as shown below).

· 12	A	42
Site	Sample type	Sample number

* Sites will be numbered according to the map shown in Figure 7.2.

* Reference codes for the materials have been chosen to conform with those in the International Project (UNECE), and are as follows:—

A	Mild steel
D	Aluminium
C	Galvanised mild steel
H	Paint system
M	Portland (limestone)
L	White Mansfield (dolomitic sandstone)
E	Copper
Y	Monk's Park (limestone)

* Numbering for the stone samples will be as follows:—

1- 3	Exposed carousel, 1 year exposure period		
4- 6	,,	2	,,
7- 9	,,	4	,,
10-12	,,	> 4	,,
13-15	Sheltered carousel, 1 year exposure period		
16-18	,,	2	,,
19-21	,,	4	,,
22-24	,,	> 4	,,
25-27	Exposed carousel, replacement at end of 1st year		
28-30	,,	,,	2nd
31-33	,,	,,	4th
34-36	Sheltered carousel, replacement at end of 1st year		
37-39	,,	,,	2nd
40-42	,,	,,	4th

* Numbering for the metal samples will be as follows:—

1- 3	1 year exposure period	
4- 6	2	,,
7- 9	4	,,
10-12	> 4	,,
13-15	Replacement at end of 1st year	
16-18	,,	2nd
19-21	,,	4th

C.5.5 Sample Mounting

A standard rack design for the mounting of materials samples has been agreed upon (Figure 7.4). Purchase of materials, and construction of the rack are the responsibility of each organisation running sites.

Racks should be located away from any obstacle which might cause wind shadow effects on open sites.

Samples and their labels will be sent in individual plastic bags in which they should be kept until the label and the samples are placed on the appropriate carousel arm or rack. Handling samples (with gloves) and labels one at a time will prevent sample identity becoming lost.

C.5.6 Monitoring procedures

The monitoring programme will involve removing THREE of each material samples after 1, 2 and 4 years. These samples will then be analysed (see Sections C.6.1 and C.6.2). The exposure periods for replacement samples will be decided at a later date.

* THREE samples of each of the minimum requirement materials will be removed after each of the three intervals and new replacement samples exposed in their place. Careful handling of samples is vital, and the numbered labels must be kept with the samples at all times.

* Photographs of samples must be taken at the end of each interval (following the guidelines set out in Sections C.5.3 (c)).

* Stone samples (with their labels) should be returned to the BRE in the plastic bags provided.

* Metal samples should be returned to the BRE.

C.6 ANALYSIS OF SAMPLES, INTERPRETATION AND TRANSMISSION OF RESULTS

A programme of analysis has been designed for stone and material samples.

C.6.1 Analysis of Stone Samples

* Stone samples will be removed, dried and cooled as before and then reweighed to investigate weight changes over the exposure periods (see Section C.5.3 (b)).

* Surface roughness, and surface elemental composition will also be investigated. More detailed investigations involving Scanning Electron Microscopy and other techniques may be used on some samples.

C.6.2 Analysis of Metal Samples

* TWO out of each group of three metal samples removed will be stripped and weighed as before (see below). Analysis of descaling solutions would be useful.

* The THIRD sample will be subjected to more detailed investigation techniques including Scanning Electron Microscopy and metallographic analysis.

(a) *Stripping procedures*

(i) Mild steel. Scrub specimens with a stiff brush in water to remove loose corrosion products. Immerse for 10 minutes in 10% H_2SO_4 (100 ml H_2SO_4, s.g. 1.84 made up to 1 litre) containing 0.5 g quinoline ethiodide. Water wash. Methylated spirit wash. Hot air dry.

(ii) Aluminium. Stripping solution; 50 ml phosphoric acid, s.g. 1.69; 20 g chromium trioxide made up to 1 litre. Immerse specimen for 5 minutes at 90 degrees C. If corrosion product film remains, immerse for 1 min in nitric acid, s.g. 1.42. Water wash, condensing solvent wash. Hot air dry.

(iii) Galvanised mild steel. Immerse specimen in acetic acid (50 ml glacial CH_3COOH made up to 1 litre) at 20 degrees C for 10 minutes. Wash in water. Condensing solvent wash. Hot air dry.

(iv) Alkyd paint. Water wash before weighing. Thickness measurement. Adhesion pull-off test. Specular gloss measurement.

(v) Copper. Immerse specimen for 30 minutes in sulphuric acid (54 ml H_2SO_4, s.g. 1.84 made up to 1 litre) at 40 degrees C. Deaerate solution with nitrogen. Brush specimens and reimmerse for 3 mins. Cold water wash. Condensing solvent wash. Hot air dry.

(b) *Weighing procedures*

(i) Samples to be weighed to 5th significant figure.

C.6.3 Interpretation and Transmission of Results

The DoE will make arrangements with individual organisations involved in the NMEP for the transmission of meteorological and pollution data to a central data bank at BRE. Summaries of the data will be made available, at not less than one year intervals, to all organisations that supply data; more detailed information will be available if specifically requested. The BERG Committee will undertake the publication of results at appropriate intervals.

D: Terms of Reference of the BERG Sub-Committees

Sub-Group I – Air Pollution Modelling

1. To review the existing work on the modelling of pollutant deposition relevant to building materials damage in the United Kingdom.

2. Where possible to draw conclusions with particular emphasis on source responsibilities.

3. To identify gaps in knowledge and to assess the need for further research.

Sub-Group II – Site Experience

1. To review the existing data acquired from studies of actual buildings and the exposure of different materials to the atmosphere.

2. To identify future requirements for site and materials exposure studies, and to advise on their implementation.

Sub-Group III – Intercomparison and Validation of Laboratory Tests Against Field Exposures

1. To review laboratory investigations of pollutant effects on the degradation of building materials.

2. To advise on suitable laboratory test methods to elucidate damage mechanisms and explain site experience.

3. To advise on the relationship between laboratory data and site exposure results.

Sub-Group IV – Applied Protection and Preservation Techniques

1. To consider the available methods and materials for protecting building materials from atmospheric attack, and for preservative treatment of materials already exposed.

2. To make recommendations for future research and development.

Sub-Group V – Materials Inventory and Evaluation of the Building Stock-at-Risk

1. To review methods for carrying out materials inventories and stock-at-risk surveys for both historical and modern buildings.

2. To review evaluation methodologies which may be used to determine costs and benefits of reduced acid deposition.

3. To examine the levels of uncertainty in inventory and evaluation methodologies.

E: Current projects

The National and International Materials Exposure Programmes (see Section 7.5) represent two major projects that are currently taking place (Summer 1988). The list below summarises some of the other current projects that are studying the effect of acid deposition on building materials. The organisations undertaking the work are given in parenthesis.

(i) Urban/Rural decay patterns reflected in limestone decay on transect from SW-NE England (UCL/BRE).

(ii) Investigation of individual buildings (some of which are part of the NMEP) eg Wells Cathedral, York Minster (BRE/UCL, CEGB).

(iii) Investigation of the relationship between stone durability and pore shape/size, and the relative durability of stone types using crystallisation tests, porosity by water absorption, and acid immersion tests. Studies on the effectiveness of stone consolidants and masonry waterproofing agents (BRE).

(iv) Methods of determining rates of decay of metals by continuous measurement (NPL, CEGB/CERL, UMIST/BCC).

(v) Investigation of historic records of pollution at specific buildings (eg Wells Cathedral, Lincoln Cathedral, York Minster). (BRE/UCL, CEGB/UEA).

(vi) Laboratory studies of the degradation of selected stones, concretes and mortars. Samples of building materials and reference samples of metals are being examined with the aid of accelerated exposure chambers and modern experimental techniques. Particular attention is being paid to NO_x pollutants; their additive or synergistic effect in the presence of SO_2 is being assessed through wet/dry cycling. (UMIST/BRE).

(vii) Investigation of the effects of acidic fogs and mists on vertical surfaces. (UMIST/BRE).

(viii) Study of methods for compiling inventories and assessing stock-at-risk. (ECOTEC/DOE/BRE).

(ix) Studies on the effect and treatment of organic growths and bacteria on stone. (BRE/Portsmouth Polytechnic).

(x) Examination of decayed stone from historic buildings. (BRE).

(xi) Modelling the dispersion of atmospheric pollutants. (WSL/DOE, CEGB, BCC).

(xii) Studies on pollutants emitted by vehicles. (WSL/DOE).

(xiii) Compilation of emission inventories for urban areas. (WSL/DOE).

(xiv) The possible influence of HCl and particulates, individually or in combination, on the degradation of stone and on the corrosion of galvanised surfaces (UMIST/BCC).

(xv) Failure examinations of components and structures exposed to atmospheric environments. Studies involve the examination of oxide jacking in steel framed buildings, concrete degradation and reinforcing bar attack, and damage to large engineering structures such as electricity transmission systems and road bridges. (UMIST/CAPCIS).

(xvi) Laboratory studies on limestone durability and the mechanical effects of expansive reactions. Laboratory methods of characterising existing damage are being assessed. (CEGB).

References – Appendix A

Armogie, C.J. and Ball, D.J. (1983). Simulation and prediction of sulphur dioxide concentrations in London's air using a computer dispersion model. Greater London Council Research Report 129. London: GLC.

Atmospheric Dispersion Modelling Working Group (1979). A model for short and medium range dispersion of radionuclides released to the atmosphere. *National Radiological Protection Board Report* NRPB-R91.

Atmospheric Dispersion Modelling Working Group (1981a). A procedure to include deposition in the model for short and medium range atmospheric dispersion of radionuclides. *National Radiological Protection Board Report* NRPB-R122.

Atmospheric Dispersion Modelling Working Group (1981b). The estimation of long range dispersion and deposition of continuous releases of radionuclides to the atmosphere. *National Radiological Protection Board Report* NRPB-R123.

Atmospheric Dispersion Modelling Working Group (1981c). A model for long range atmospheric dispersion of radionuclides released over a short period. *National Radiological Protection Board Report* NRPB-R124.

Atmospheric Dispersion Modelling Working Group (1983). Models to allow for the effects of coastal sites, plume rise, and buildings on dispersion of radionuclides and guidance on the value of deposition velocity and washout coefficients. *National Radiological Protection Board Report* NRPB-R157.

Ball, D.J. (1980). Estimation of the atmospheric dispersion of pollution in urban and suburban areas. *GLC Scientific Branch Report* R90.

Ball, D.J. and Amorgie, C.J. (1980). Implications for London of European air quality standards for sulphur dioxide and suspended particulates. *Environmental Pollution* (B), 5, 207-232.

Ball, D.J. and Radcliffe, S.W. (1979). An inventory of sulphur dioxide emissions to London's air. *Greater London Council Research Report 23*. London: GLC.

Barker, C.D. (1978). A comparison of Gaussian and diffusivity models of atmospheric dispersion. Central Electricity Research Laboratory Report RD/B/N4405.

Bennett, M., Rogers, C. and Sutton, S. (1986). Mobile measurements of winter SO_2 levels in London 1983-4. *Atmos.Environ.*, 20, 461-470.

Briggs, G.A. (1969). *Plume Rise.* US Atomic Energy Commission.

Briggs, G.A. and Moore, D.J. (1975). A comparison of the trajectories of rising buoyant plumes with theoretical empirical models *Atmos.Environ.*, 9, 455-459.

Digest of Environmental Protection and Water Statistics, No.7, Part 2, Air Pollution (1984).

Fisher, B.E.A. (1975). The long range transport of sulphur dioxide. *Atmos.Environ.*, 9, 1063-1070.

Fisher, B.E.A. (1978). The calculation of long term sulphur deposition in Europe. *Atmos.Environ.*, 12, 489-501.

Fisher, B.E.A. (1983). A review of the processes and models of long-range transportation of air pollutants. *Atmos.Environ.*, 17, 1865-1880.

Garland, J.A. (1978). Dry and wet removal of sulphur from the atmosphere. *Atmos.Environ.*, 12, 349-362.

Gifford, F.A. and Hanna, S.R. (1973). Urban air pollution modelling. In proceedings of the 2nd International Clean Air Congress (IUAPPA). 1146-1151. Academic Press.

Kallend, A.S. *et al* (1983). The fate of atmospheric emissions along plume trajectories over the North Sea. Vol. 3, Modelling Studies. *Central Electricity Research Laboratory Report TPRD/L/R82.*

Keddie, A.W.C., Bower, J.S., Maughan, R.A., Roberts, G.H. and Williams, S.P. (1978). The measurement, assessment and prediction of air pollution in the Forth Valley of Scotland: Final report. *Warren Springs Laboratory Report* LR 279(AP).

Moore, D.J. (1974). A comparison of the trajectories of rising buoyant plumes with the theoretical/empirical models. *Atmos.Environ.*, 8, 441-457.

Moore, D.J. and Lee, B.Y. (1982). An asymmetrical Gaussian plume model. *Central Electricity Research Laboratory Report* RD/L/2224/N81.

Organisation for Economic Co-operation and Development (OECD) (1984). *International comparison study on reactor accident consequence modelling.*

Perrin, D.A. (1986). Modelling the transport and removal of sulphur dioxide emissions in the United Kingdom. *Warren Spring Laboratory Report* LR 560 (AP)M.

Review Group on Acid Rain (RGAR) (1983). *Acid Deposition in the United Kingdom.* London: Warren Spring Laboratory.

Rodhe, H. and Grandell, J. (1972). On the removal time of aerosol particles from the atmosphere by precipitation scavenging. *Tellus*, 24, 442-454.

Rote, D.M. (1980). Gaussian plume models. In *Atmospheric Planetary Boundary Layer Physics* (Longhetto, A. Ed). Elsevier.

Scriven, R.A. and Fisher, B.E.A. (1975a). The long range transport of airborne material and its removal by deposition and washout - I. General considerations. *Atmos.Environ.*, 9, 49-58.

Scriven, R.A. and Fisher, B.E.A. (1975b). The long range transport of airborne material and its removal by deposition material and washout - II. The effect of turbulent diffusion. *Atmos.Environ.*, 9, 59-68.

Singer, I.A. and Smith, M.E. (1966). An improved method of estimating concentrations and related phenomena from a point source emission. *J.Appl.Meteorol.*, 5, 631-639.

Smith, F.B. (1973). A scheme for estimating the vertical dispersion of a plume from a source near ground level. Unpub. Meteorological Office Note.

Smith, F.B. (1981). The significance of wet and dry synoptic regions for long-range transport of pollution and its deposition. *Atmos.Environ.*, 5, 863-873.

Spanton, A.M. (1983). Dispersion modelling of sulphur dioxide concentrations in London with reference to the EC Directive Limit Values. *Warren Spring Laboratory Report* LR 457(AP).

References – Appendix B

Larssen, R.I. (1969). A new mathematical model of air pollution concentration averaging time and frequency. *JAPCA*, 19, 24-30.

Simpson, R.W., Butt, J. and Jakeman, A.J. (1984). An averaging time model of SO_2 frequency distributions from a single point source. *Atmos.Environ.*, 18, 1115-1124.

Printed in the United Kingdom for Her Majesty's Stationery Office
Dd. 291302 C15 1/89